Cellulite:
Defeat It Through Diet and Exercise

The Vanguard Press, Inc.   New York

CELLULITE
Defeat It Through Diet and Exercise

Beverly Cox & George Benois

photos by Mike Yamaoka

The authors would like to acknowledge the assistance of Harriet Reilly, Catherine Elish, Toni Fitch, Joan Toppan, and the Richard Ginori Corporation of America.

Copyright © 1981 by Beverly Cox and George Benois
Published by Vanguard Press, Inc.,
424 Madison Avenue, New York, N.Y. 10017.

No part of this publication may be reproduced or transmitted in any form or by any means, electronic or mechanical, including photocopy, recording, or any information or retrieval system, or otherwise, without the written permission of the publisher, except by a reviewer who may wish to quote brief passages in connection with a review for a newspaper, magazine, radio, or television.

Designer: Elizabeth Woll

Manufactured in the United States of America

1   2   3   4   5   6   7   8   9   0

Library of Congress Cataloging in Publication Data

Cox, Beverly, 1945–
  Cellulite: defeat it through diet and exercise.

  Includes index.
  1. Reducing diets — Recipes. 2. Reducing exercises. I. Benois, George. II. Title.
  RM222.2.C65   616.3'9806   81-1735
  ISBN 0-8149-0845-4   AACR2
  ISBN 0-8149-0846-2 (pbk.)

Contents

FOREWORD   9

ABOUT THE PROGRAM   17

PROGRAM DAY ONE   27
  RECIPES
    Omelet with Herbs   28
    Grated Carrot and Orange Juice Salad   29
    Harira (Hearty Moroccan-Style Soup)   30
    Yogurt with Cinnamon and Honey   32
    Fennel and Arugula Salad   33
    Grilled Salmon Steaks with Yogurt Béarnaise
      Sauce   34
    Strawberry Ice   36
  EXERCISES
    Starting Position   37
    Knee to Chest Position   38
    Leg Lift-ups   39
    Pelvic Lift-up   40
    Ball-up Position   41

CONTENTS

PROGRAM DAY TWO   43
   RECIPES
      Mushroom, Watercress, and Endive Salad   44
      Individual Cheese Soufflés with Herbs   45
      Frozen Peach Yogurt Melba   47
      Oriental Watercress Soup   48
      Southeast Asian-Style Beef Kabobs Served in Lettuce and Herb Wrappings with Peanut Sauce   49
      Steamed Honey Pears   51
   EXERCISES
      Spine Rock   53
      Off-balance Sitting Position   54
      Frog Balance   56
      Tent Position   57

PROGRAM DAY THREE   59
   RECIPES
      Banana or Peach Energy Drink   60
      Orange and Onion Salad   61
      Puffy Asparagus and Cheese Omelet   62
      Louisiana Honey Custard   64
      Chilled Curried Fruit Soup   66
      Breast of Chicken Tandoori   68
      Kitcheri (Brown Rice Cooked with Lentils)   70
      Ananas Glacé (Pineapple Sherbet Served in a Pineapple)   72
   EXERCISES
      Shoulder Stand   74
      Pelvic Lift-ups   75
      Back Leg Lifts   76

CONTENTS

PROGRAM DAY FOUR 79
- RECIPES
  - Yucatecan Chicken and Lime Soup 80
  - Strawberries in Orange Baskets 82
  - Cold Cucumber-Yogurt Soup 84
  - Baked Fish Steaks Moroccan Style 86
  - Papaya Boats 88
- EXERCISES
  - Side Leg Lifts 90
  - Back Leg Lifts 92
  - Knee Press-downs 93
  - Forward Stretch 94

PROGRAM DAY FIVE 97
- RECIPES
  - Huevos Rancheros (Baked Eggs Mexican Style) 98
  - Exotic Fruit Salad with Vinaigrette Sauce 100
  - Banana Sherbet 102
  - Green Bean and Onion Salad 104
  - Chicken Breasts with Sweet and Sour Sauce, Bouquet of Vegetables, and Tart Apple Purée 106
  - Lemon Soufflé 109
- EXERCISES
  - Cobra Position 111
  - Bow Posture 112
  - V Posture 113
  - Cat Hump Resistance Stretch 114
  - Camel Posture 116
  - Plow Posture 117

CONTENTS

PROGRAM DAY SIX   119
  RECIPES
    Eggs Parmentière (Eggs Baked in Potatoes) with Herbed
        Yogurt Hollandaise Sauce   120
    Melon Swan Filled with Fruit   124
    Cold Asparagus with Vinaigrette Sauce   126
    Lamb Chops Provençal   128
    Tomatoes Provençal   130
    Cold Strawberry Yogurt Soufflé   132
  EXERCISES
    Lotus Postures   134
    Standing Stretches   136

PROGRAM DAY SEVEN   139
  RECIPES
    Soft-Boiled Egg with Asparagus Tips   140
    Sole Filets with Julienne of Vegetables Baked in
        Papillotes   142
    Ricotta Pudding   145
    Gaspacho Andaluz (Spicy Cold Tomato Soup)   146
    Lemon Chicken Brochettes   148
    Brown Rice Pilaf   150
    Oranges with Cinnamon   152
  EXERCISES
    Standing Chest Expansion   154
    Triangle Twist   155
    Back Bend and Twist   156
    Hip Balance   157

INDEX   158

Foreword

Defeat Cellulite through Diet

ANY WOMAN WHO has suffered through starvation diets and strenuous exercise trying, to no avail, to rid herself of dimpled thighs and saddlebag hips knows that there must be a difference between ordinary fat and cellulite.

Cellulite is the Number One figure problem for approximately 80% of American women over the age of twelve. The lumpy cottage-cheese-like patches that are characteristic of the condition are usually found on hips, buttocks, thighs, and legs, but are also occasionally present on the abdomen, back, and upper arms.

Cellulite is sometimes referred to as "thin women's fat" because it afflicts thin, physically active women as well as those who are overweight. Even professional dancers, models, and athletes are not immune to its ravages.

The concept that cellulite is different from ordinary smooth fat is still quite new in the United States. In Europe, however, it has been recognized as a separate condition since the early 1900s. Swedish scientists were the first to "discover" the condition and prescribe regimes of massage and special diet to combat it. The French, always front runners in treating feminine beauty problems, coined the term "cellulite" and soon became expert in its treatment.

According to cellulite authorities, normal smooth fat is the body's storehouse for excess energy. When physical activity is increased or calorie intake is decreased, the body calls upon this energy reserve and fat is reduced.

FOREWORD

Cellulite, on the other hand, is thought to be the result of congestion of the layer of connnective tissue that lies between the skin and the muscles and internal organs. This connective tissue plays an important role in the proper function of the body because, among its components, are the circulatory vessels and lymphatic fluid that carry nourishment to the cells and carry back wastes for elimination. When, because of sluggish body function, wastes cannot be efficiently eliminated, a vicious circle begins. Wastes engorge surrounding tissues and harden into lumpy pockets. These pockets further impair circulation in the area and more pockets form.

Another unpleasant effect of this congestion is that it keeps nourishment from reaching the cells. As a result of lack of nourishment skin is often dry, sensitive, and prone to bruising and broken blood vessels. Varicose veins also worsen when circulation is impaired by cellulite.

Since 90% of all cellulite cases occur in women and appear to be triggered by periods of hormonal change such as puberty, pregnancy, use of birth-control pills, and menopause, it is strongly suspected that the female hormone estrogen is a contributing factor in cellulite formation. A hereditary tendency toward cellulite is also possible. However, there is no reason to despair if "fat thighs run in the family." Poor nutrition, insufficient water intake, tension, fatigue, and lack of consistent exercise are also major contributors to the condition, and all can be controlled.

European doctors and nutritionists usually prescribe a cleansing regime that combines an additive-free and controlled sodium diet with a consistent, but unstrenuous, exercise program. The result is improved digestion, elimination, circulation, and body tone. New cellulite formation is discouraged and eventually old formations break down.

Sometimes medical experts work in conjunction with specialized beauty institutes that offer massage and other treatments that encourage quicker cellulite dissipation. But even with beneficial extras like massage, cellulite control must be viewed as a long-term program. Unlike five pounds gained over the holidays, your cellulite did not appear in a week nor will it disappear with a few days of dieting. Crash dieting and strenuous exercise can actually aggravate a cellulite condition because they shock and strain an already overtaxed system. In fact, unless you are overweight, stop thinking calories and think good nutrition!

I first became interested in cellulite and its treatment during the five years I spent in France studying cooking and working as a model. At five feet nine inches and 120 pounds I was not overweight, but I had the beginnings of a cellulite condition.

The hardest part of following my doctor's prescribed regime was the diet. To someone with an interest in good food, it seemed very Spartan. The French are fond of saying "Il faut souffrir pour être belle" (you must suffer to be beautiful) and this diet convinced me it was true. No salt, no sugar, no alcohol — not even wine — and apparently few appealing alternatives. Consequently, though I was concerned about those thigh bulges, especially during bikini season, I found the treatment more painful than the complaint. I could follow the diet without cheating for short periods of time, but I could not accept it as a way of life. It was simply too boring!

Doctors and nutritionists make a valuable contribution to our health by offering proper dietary advice. However, it is unrealistic to expect them to make dieting a pleasant culinary experience as well.

Reflecting on my own cellulite experience, it occurred to me that many people on special diets could use some help from a cook. In *Gourmet Minceur: A Week of Slimming Cuisine*

FOREWORD

and in *Minceur Italienne,* a collection of slimming Italian menus with recipes, I confronted the problem of making dieting more pleasant for those with a weight problem. Therefore, it seemed possible that, with a little imagination, a book of appealing anti-cellulite recipes could be developed.

At that time I had already begun to work with George Benois and found that my body responded well to the gentle coaxing of Yoga. My circulation and body tone had greatly improved and I felt I could now hold my own in the war against cellulite. What I needed to defeat it was a well-planned series of good-tasting recipes that would allow me to stick to the proper diet.

Before beginning to develop these recipes, I consulted the diet suggestions of cellulite experts, both in Europe and the United States. My goal was to arrive at a consensus of basic "do's" and "don'ts" in combating cellulite, along with a list of the foods most beneficial toward that end. I also had to take into consideration that many women are not only cooking for themselves, but also for their families.

Fortunately, I found that an effective anti-cellulite diet is not only nutritious, but can also be delicious. My biggest challenge was to develop recipes without added salt that would please even a salt lover's palate. For inspiration, I looked to a variety of international cuisines in which herbs and spices are used skillfully and with interesting results.

There are numerous delicious foods that are also cellulite fighters — among them strawberries, seafood, asparagus, tomatoes, artichokes, and even the much-maligned potato. Salt intake can be painlessly limited by seasoning with lemon or lime juice, onions, fresh herbs and spices. Pure unfiltered "raw honey" is a satisfying substitute for sugar.

The "Defeat Cellulite" program is *not* a weight-loss diet, but rather part of a purifying regime designed to rid the body of toxins and encourage the breakdown of cellulite. However, along with sodium counts I have included, per portion, calorie and carbohydrate counts for those who also have a weight problem. Most recipes serve four, but can be halved or doubled. The suggested portion size should satisfy the average woman's appetite, but portions can be adjusted for men and children who can also benefit from less salt and fewer additives.

I hope you will enjoy these recipes and find the seasoning suggestions and cellulite-defeating tips helpful. But remember, it is always a good idea to check with your doctor before starting any new eating or exercise regime.

There is no reason why a bikini figure cannot co-exist with a satisfied palate!

Beverly Cox

FOREWORD

Defeat Cellulite through Exercise

WE HAVE COME to realize that taking responsibility for our own health is vital. The body is a magnificent machine; keeping it well toned and properly nourished is our best possible insurance against malfunction and disease.

Cellulite is a good example of body malfunction that can be remedied by proper nutrition and exercise. Three of the major causes of cellulite formation are tension, sluggish elimination, and poor circulation — all common hazards in our modern society.

Yoga is unique in treating cellulite. It is the only system of physical-mental culture that acts directly upon the endocrine (glandular) system, increasing the secretions and tone; that brings the system into greater balance and harmony. Yoga exercises are also effective in preventing sluggish elimination, circulatory interference, and glandular overexcitements and imbalances. These exercises limber the joints, stretch the spine, squeeze and stretch muscles and ligaments. Through this form of self-massage, the system is stimulated and cleansed. Unhealthy fat deposits that inhibit proper circulation are broken down and flushed away. Eventually, the body is remolded to its rightful contours.

The Yoga exercises in the Defeat Cellulite regime have been tested in my teaching laboratory over many years. While the following exercises do not constitute a complete course in Yoga, they do represent and may be used as a foundation and clarification of its principles. The movements are preventive and curative for the condition we are dealing with, and are remarkable revitalizers for the whole system. Hundreds of students have realized positive benefits. No program is magic, however; the secret of success is to maintain a consistent schedule of exercise, not just for one week but for a lifetime. After all, we do not stop brushing our teeth after our teen-age years. Why then should the body, our vehicle for life, be subjected to inactivity? I do not imply that you must become a health fanatic. Even a few minutes a day devoted to this unstrenuous but well-planned exercise program will bring worthwhile results.

Strive for a youthful body that has flexibility, lightness, and elasticity — a spine that bends forward, backward, and twists! Be enthusiastic and disciplined! Start today, and whether you are fifteen or fifty, you will look and feel better in ten years than you do now.

George Benois

About the Program

Menus and Recipes

THE KEYS TO success in cellulite control are moderation, balance, and harmony!

Consuming large amounts of even the most beneficial foods at one sitting may put a strain on the system and cause bloating. So don't be tempted to gorge yourself on strawberries or gulp down two or three glasses of water. Individual tolerance to even moderate amounts of certain foods may vary. For example, some women find it difficult to digest even small amounts of raw vegetables. If this is your case, you will probably find lightly steamed vegetables more digestible until your system begins to work more efficiently.

FRESH FRUITS AND VEGETABLES

Fresh fruits and vegetables should play an important part in your diet. The ones listed below are considered especially beneficial in controlling cellulite.

apples	grapefruit	radishes
asparagus	leeks	rhubarb
artichokes	lettuce	spinach
bananas	onions	strawberries
cabbage	papaya	string beans
carrots	pears	tomatoes
fennel	pineapple	turnips
garlic	potatoes	watercress

ABOUT THE PROGRAM

OTHER HIGHLY RECOMMENDED FOODS AND SUPPLEMENTS

plain yogurt that contains active cultures
egg yolks—except for those on low-cholesterol diets
brown rice
legumes (lentils, chickpeas, dried beans, etc.)
unsalted nuts
sunflower, sesame, and pumpkin seeds
unsweetened wheat germ (regular or toasted)
brewer's yeast

FISH, POULTRY, AND MEAT

Lean fish, poultry, and small amounts of meat are good sources of protein. Fat and skin should not be eaten. Roasting, baking, broiling, and sautéing with little or no oil in a non-stick skillet are the best methods of preparation.

CHEESE

There are a variety of unsalted cheeses available in cheese shops and health food stores. My favorites are Swiss Lace and salt-free Cheddar. Other good low-fat and low-salt cheeses are Jarlsberg, Gouda, Dambo, and Fontina. Among the pot cheeses, low-sodium cottage cheese, farmer cheese and ricotta are good choices. Up to 3 ounces of these cheeses may be eaten each day.

OIL AND BUTTER

Up to 2 tablespoons of pure *cold pressed* vegetable oils (so indicated on the label, usually found in health-food stores) are allowed per day. Safflower, peanut, sesame, and olive are good choices.

Unsalted butter is limited to 1 teaspoon per day. Whipped butter offers more volume for fewer calories.

BREAD AND CEREALS

If you are not overweight, bread can be eaten in moderation. Gluten bread, whole wheat pita, and whole grain breads without chemical additives or preservatives are good choices.

Reasonable portions of cereals without refined sugar or preservatives may be included occasionally in your diet. Some of the good ones are Familia, granola sweetened with honey, bran, and shredded wheat.

SWEETENERS

Unrefined honey (labeled "raw" honey) has won the endorsement of experts such as cellulite authority Nicole Rosard and famous nutritionist Paavo Airola as an ideal sweetener.

Unrefined honey contains small amounts of minerals, B-complex vitamins, and vitamins C, D, and E. It is easily digestible and is thought to be beneficial in cases of poor circulation and constipation.

WATER

Drinking enough water is essential to combat cellulite. Start with 4 and gradually build up to 6 or even 8 glasses per day. Drink small amounts at a time to avoid bloating. And *never* drink water or other beverages during meals.

ABOUT THE PROGRAM

HERBAL TEAS

Whenever possible, drink herbal teas instead of coffee or tea. There are many good ones available, but my favorite is the natural diuretic mixture served by famous diet chef Michel Guérard at his spa-hotel at Eugénie-les-Bains in southwest France. Tisane d'Eugénie contains equal parts of horsetail, heather flowers, cherry stems, corn silk, and bearberry. If your health food store does not stock all these ingredients, look for teas that contain at least some of them. Except at breakfast when a little cheating is allowed, herbal tea should be consumed at the end of meals.

SEASONINGS

Most herbs, fresh and dried, and many spices like cumin, curry powder, sweet paprika, nutmeg, cinnamon, ginger, and turmeric are good seasoning choices. Cayenne pepper is controversial, but some nutritionists, among them Paavo Airola, feel that it aids digestion.

Black pepper and Dijon mustard (low-sodium if possible) may be used occasionally with great moderation. Lemon and lime juice, onions and garlic may be used freely.

FOODS TO AVOID

Salt, sugar, and white flour
All canned, frozen, packaged, processed, and preserved foods containing chemical additives, sodium (salt), sugar, or white flour
Poultry and meats with a high fat content such as duck, goose, pork, and all pork products
Fried foods

ABOUT THE PROGRAM

Alcohol, including beer and wine
Carbonated beverages

ABOUT THE INGREDIENTS

Unless otherwise specified, all fruit and vegetables are fresh and of average size.

Eggs are large.

Vegetable oils should be cold pressed.

Honey is the natural, raw, unfiltered kind found in health food stores.

SUBSTITUTIONS IN MENUS

My goal in this series of menus is to provide a variety of recipe suggestions. Menus may be simplified by using some of the following substitutions:

A raw vegetable salad or a plate of lightly steamed vegetables eaten alone or with a 4 oz. serving of lean roasted or broiled fish, poultry, or meat may be substituted for lunches or dinners. Poultry and meat consumption should, however, be limited to one meal a day.

Fresh fruit, 2 oz. servings of cheese (for suggested list, see page 18), or 4 Tb of plain yogurt with wheat germ and honey are good substitutions for other desserts.

Two Simple Basic Recipes

Low-Sodium Chicken or Turkey Broth

MAKES 3 QT.
½ CUP SERVING CONTAINS APPROXIMATELY 15 CALORIES •
1 GR. CARBOHYDRATES • 10 MG. SODIUM

Canned chicken broth and beef broth, unless specifically marked "low sodium," are not only very salty but also contain additives such as monosodium glutamate that are detrimental to an anti-cellulite regime.

The following recipe is simple to make and a good way to use up wing tips, necks, and giblets as well as leftover carcasses of roast fowl. I usually freeze these pieces until there is enough for a batch of broth and sometimes supplement them with inexpensive parts like backs. The flavor of the broth can be varied by adding other vegetables such as turnips, leeks, or parsnips that you may have on hand. It is also interesting to experiment with other fresh or dried herbs like sweet basil, tarragon, chervil, or oregano.

ABOUT THE PROGRAM

INGREDIENTS

3 to 4 lb. chicken or turkey bones. At least 1½ lb. should be uncooked necks, backs, wingtips, or giblets
6 carrots, washed but not peeled, cut into large chunks
3 onions, peeled, quartered, and stuck with 8 cloves
5 branches celery with some leaves, washed and cut into large chunks
1 tsp coarsely ground black pepper or ½ tsp whole black peppercorns
2 bay leaves
1 tsp dry thyme or 5 branches fresh thyme
½ cup well-washed parsley leaves and stems
5 qt. cold water to cover

METHOD

1. Place bones in a large stockpot or Dutch oven. Add cold water to cover and bring to a boil, skimming off any foam that rises to the top.

2. Add vegetables and seasonings. Return to a boil, then lower heat and simmer partially covered for 1 hour.

3. Uncover and continue to simmer until liquid has reduced to approximately ½ the original volume, about 1 hour.

4. Strain broth through a sieve lined with cheesecloth, pressing on bones and vegetables to extract juice.

5. Skim off any fat that has risen to the top of the broth. This is easier to do if broth has been chilled or frozen. When freezing, make sure to leave room in containers for expansion.

ABOUT THE PROGRAM

Gluten Bread with Variations

MAKES 1 LOAF

PER SLICE:

Gluten Bread	86 Calories	10.5 gr. Carbohydrates	7.5 mg. Sodium
Gluten Triticale	80 Calories	12 gr. Carbohydrates	7.5 mg. Sodium
Gluten Triticale and Wheat Germ	65 Calories	12 gr. Carbohydrates	7.5 mg. Sodium

Gluten is the high-protein, low-starch part of wheat. Because of these qualities gluten bread is often recommended in cellulite-control diets.

Though gluten bread without preservatives is available in health food stores, I thought it might be fun to include a simple recipe for anyone who enjoys making bread.

Triticale is a high-protein hybrid of red wheat and rye. Triticale flour mixed with gluten flour or with gluten flour and wheat germ makes a darker, heartier loaf but is not as high in protein.

INGREDIENTS

½ Tb (½ package) dry yeast
1 tsp honey
½ cup warm water
1 cup warm milk (made with 6 Tb non-fat dry milk)
¼ tsp salt (optional)
1 tsp honey
1 Tb melted unsalted butter
3 cups gluten flour, or 2 cups gluten and 1 cup triticale flour or 1¾ cups gluten
2 tsp safflower oil

METHOD

1. Combine yeast, honey, and warm water and let it stand for 5 minutes to dissolve the yeast.

2. Add the remaining ingredients in the order listed and stir with a wooden spatula until well combined.

3. Turn dough onto a floured surface and knead with lightly oiled hands for 5 minutes. To knead, push dough away from you with the heel of one hand while holding on to it with the other hand. Turn the dough a quarter turn after each kneading motion.

4. Place dough in a lightly oiled bowl. Lightly oil the top of dough to keep it from crusting and cover bowl with waxed paper or a dish towel.

5. Let dough rise in a warm, draft-free area until it has doubled in size. Approximately 1½ hours.

6. Lightly oil hands and knead dough for 3 to 5 minutes more and place in oiled loaf pan (approximately 5" by 9").

7. Lightly oil the top of dough and cover. Let rise for 1 hour or until it has filled the pan and risen slightly above the sides.

8. Bake in the middle of a preheated 350° oven for 50 minutes or until loaf is nicely browned and slightly crisp on the outside.

9. Unmold and cool bread on a wire rack. Store in the refrigerator in a plastic bag.

ABOUT THE PROGRAM

The Exercises

THE EXERCISES ILLUSTRATED on the following pages have all proved helpful in treating your cellulite condition and, along with proper diet, eliminating it. As a whole, they are a general guide to the control of cellulite build-up and, with diet, some of them should become an integral part of your daily routine (eventually you can create your own set of exercises, choosing those most directed to your specific needs — be it the back, thighs, arms, or wherever). A diversity of exercises repeated a few times is preferable to selecting only a few and repeating them often.

All the movements are self-applied manipulations. You will find that some can be effected immediately; those that you cannot thus adapt to completely should be performed only as far as the movement can be executed at the moment. Never force your body beyond that moment's capability: by gentle coaxing and encouraging the effort, in time your body will mold itself into the exercise's proper final position.

In addition to their specific purpose, the exercises stimulate the circulation, relieve tension, and counteract the ill effects of poor posture — all factors in the build-up of cellulite. Always be aware of your posture, trying to move with ease and grace — keeping your spine straight, stomach muscles pulled in and up, and shoulders back and relaxed.

Respect for your body will bring its own rewards — a feeling of greater achievement, more energy, better health . . . and less cellulite.

Program Day One

Menu

BREAKFAST

    1 slice fresh pineapple or ½ grapefruit
    Omelet with herbs
    1 slice gluten toast (optional)
    ½ tsp whipped unsalted butter (optional)
    1 tsp honey (optional)
    Herb tea

LUNCH

    Grated carrot and orange juice salad
    Harira (hearty Moroccan-style soup)
    Yogurt with cinnamon and honey
    Herb tea

DINNER

    Fennel and arugula salad
    Grilled salmon steaks with yogurt béarnaise sauce
    Strawberry ice
    Herb tea

Exercises

- STARTING POSITION
- KNEE TO CHEST POSITION
- LEG LIFT-UPS
- PELVIC LIFT-UP
- BALL-UP POSITION

Omelet with Herbs

SERVES 1

96 CALORIES • 5 GR. CARBOHYDRATES • 70 MG. SODIUM

INGREDIENTS

1 egg
1 tsp minced fresh basil or ⅛ tsp dried sweet basil
1 tsp minced fresh parsley
1 tsp fresh snipped chives (cut with scissors)
¼ tsp lemon juice
Pepper to taste
½ tsp unsalted butter (optional)

METHOD

1. Beat egg with a whisk or fork until white and yolk are well mixed.

2. Stir in herbs and seasonings.

3. Melt butter in a small non-stick skillet or omelet pan, spreading it evenly over the pan's surface.

4. Place pan over medium heat and add omelet mixture.

5. Quickly stir the omelet twice, with a wooden or plastic spatula, bringing cooked portions toward the center of the pan and allowing the liquid to flow toward its edges.

6. Tap the bottom of the pan once against the heating element to smooth the bottom of the omelet. Fold still moist omelet in half or roll it.

7. Allow the bottom of folded omelet to brown for a few seconds if desired, but be careful not to overcook it.

8. Serve at once.

Grated Carrot and Orange Juice Salad

57 CALORIES • 12.9 GR. CARBOHYDRATES • 47.25 MG. SODIUM

INGREDIENTS

4 carrots
½ cup freshly squeezed orange juice
1 tsp lemon juice
4 large crisp lettuce leaves (washed and dried)
Black pepper, and cinnamon or cumin to taste

GARNISH:
4 sprigs of parsley

METHOD

1. Wash, peel, and grate carrots.

2. Toss with orange and lemon juice.

3. Season to taste with pepper and cinnamon or cumin and allow to marinate refrigerated until ready to serve.

4. Arrange lettuce leaves on 4 salad plates and mound carrot salad on top.

5. Serve slightly chilled, garnished with parsley sprigs.

PROGRAM DAY ONE

Harira (Hearty Moroccan-Style Soup)

305 CALORIES • 23.2 GR. CARBOHYDRATES • 127.0 MG. SODIUM

INGREDIENTS

¼ cup chickpeas that have been soaked overnight in cold water (or ½ cup canned chickpeas well rinsed to remove salt)
1 tsp safflower oil
¾ lb. lean lamb stew meat, including some bones, cut into 1-inch cubes
1 cup minced onion
½ cup chopped celery
¼ cup minced parsley
½ tsp ground black pepper
½ tsp turmeric
½ tsp cinnamon
¼ tsp ground ginger
Pinch saffron (optional)
1 lb. ripe tomatoes, peeled and chopped, or 1 16-oz. can low-sodium tomatoes, chopped
¼ cup lentils, well washed
6 cups water
8 peeled pearl onions
1 large egg beaten with 2 Tb lemon juice

GARNISH:
4 lemon slices
Ground cinnamon

METHOD

1. Cover a large non-stick skillet with a film of the safflower oil.

2. Sauté lamb, onion, celery, parsley, and spices over medium heat for 5 minutes. Add tomatoes and cook for 10 minutes, stirring frequently.

3. Transfer lamb and vegetables to a 3- or 4-qt. heavy saucepan with lid, or Dutch oven. Add water, lentils, and chickpeas, unless they are canned.

4. Bring to a boil, then lower heat and simmer partially covered for 1½ hours.

5. With a small knife cut a cross in the root end of each pearl onion to assure even cooking, then add onions to soup and cook for ½ hour more. If using canned chickpeas, add along with onions.

6. Just before serving, turn off heat and add egg and lemon juice to soup and stir.

7. Taste and adjust seasoning, adding more lemon juice and pepper if desired.

8. Serve garnished with lemon slices and sprinkled with cinnamon. Accompany with toasted individual whole-wheat pita breads.

PROGRAM DAY ONE

Yogurt with Cinnamon and Honey

85 Calories • 11.5 gr. Carbohydrates • 26 mg. Sodium

INGREDIENTS

1⅓ cups plain yogurt
4 tsp honey
4 tsp toasted unsweetened wheat germ (optional)
cinnamon to taste

Garnish:
4 strawberries

METHOD

1. Combine yogurt, honey, and wheat germ.

2. Spoon into chilled stemmed wine glasses. Sprinkle with cinnamon and top with a strawberry.

Fennel and Arugula Salad

83 CALORIES • 7.5 GR. CARBOHYDRATES • 20 MG. SODIUM

INGREDIENTS

2 large bulbs fresh fennel
1½ cups arugula leaves, washed and dried with paper towels
1½ Tb olive oil
1¼ tsp wine vinegar
Freshly ground pepper to taste

NOTE: If fennel is not available, celery may be substituted and a few drops of anise-flavored liqueur added to dressing.

METHOD

1. Cut off stalks of fennel and reserve for other use.

2. Peel off the stringy outside of the bulb with a vegetable peeler or sharp knife.

3. Cut the bulb in quarters and core. Then slice quarters into thin crescents. Wash in cold water and dry with paper towels.

4. Combine arugula and fennel crescents in a bowl and add oil, vinegar, and pepper to taste.

5. Toss gently to coat all ingredients lightly. Taste and adjust seasonings.

6. Arrange salad attractively on 4 chilled salad plates.

PROGRAM DAY ONE

Grilled Salmon Steaks with Yogurt Béarnaise Sauce

	Calories	gr. Carbohydrates	mg. of Sodium
Salmon	300.0	6.5	185.0
Sauce	47.5	3.65	13.25
Pea Pods	25.0	6.46	13.25
Total per Serving	372.5	16.61	198.35

INGREDIENTS

1⅓ lb. fresh salmon — 4 steaks approximately ⅓-inch thick (halibut is a good substitute)

⅓ cup fresh lime or lemon juice

1 Tb safflower, corn, or peanut oil

Dash of paprika

Sauce:

2 Tb wine vinegar

2 tsp minced shallot

2 tsp minced fresh tarragon or ¼ tsp dried

1 large egg yolk

¾ tsp cornstarch

⅓ cup plain low-fat yogurt

1 tsp Dijon mustard

2 tsp sweet butter

Garnish:

½ lb. fresh sugar snap or snow peas (washed and strings removed)

4 lemon or lime wedges

METHOD

1. Place fish in a shallow dish and marinate in lime juice 30 minutes or longer, turning occasionally.

2. While salmon marinates, prepare the sauce.

3. Place vinegar, shallot, and tarragon in the top of a double boiler (or a small saucepan that will fit into another). Place over direct medium heat and reduce until almost all vinegar has evaporated.

4. Off heat, add egg yolk and cornstarch to vinegar reduction and whisk until well mixed, using a wire whisk.

5. Fill the bottom of a double boiler with water and bring to a simmer.

6. Cook the sauce over the simmering water, stirring constantly with a wooden spatula until it begins to thicken slightly, approximately 1 minute.

7. Off heat, whisk yogurt and mustard into the sauce.

8. Watching carefully that the water does not come to a boil, continue to cook sauce, stirring constantly for 1 minute.

9. Whisk butter into warm sauce.

10. Sauce may be kept warm over warm water for at least 1 hour, whisking occasionally.

11. Preheat broiler or prepare grill.

12. 15 minutes before serving, brush salmon lightly with oil, sprinkle with paprika, and cook 5 to 7 minutes on each side or until fish is firm and flaky.

13. While fish is cooking, steam peas very lightly, not more than 1 minute.

14. Place salmon on warm plates. Spoon over sauce and garnish with peas and lemon wedges. Serve at once.

PROGRAM DAY ONE

Strawberry Ice

SERVES 6

54 Calories • 13.5 gr. Carbohydrates • 2 mg. Sodium

INGREDIENTS

1 pt. fresh ripe strawberries, hulled
3 Tb orange juice
2 Tb lemon juice
3 Tb honey
1 cup water

Garnish:
6 whole strawberries
Fresh mint

METHOD

1. Purée hulled strawberries.

2. Stir in orange juice, lemon juice, honey, and water (if not sweet enough for your taste, add more honey (61 calories, 16.5 gr. carbohydrates, 1 mg. sodium per Tb).

3. Freeze ice in an ice-cream freezer, following your machine's directions for sherbets.

4. Serve ice in chilled stemmed wine glasses. Garnish with whole strawberries and mint leaves.

Note: Off-season, frozen unsweetened strawberries may be used. Purée in a food processor with above ingredients except for water. The result is an instant ice.

PROGRAM DAY ONE

I. STARTING POSITION

Lie down, keeping knees up, heels apart on floor, and hands at side.

II. KNEE TO CHEST POSITION
for abdomen and legs

Beginning in starting position, clasp hands around or just above ankle; bring knee to chest and nose to knee, hold 4 counts, release and reverse legs.

III. LEG LIFT-UPS
for abdomen, legs, and back

(a) ONE LEG LIFT-UP

Beginning in starting position, extend left leg straight to floor; lift straight up and hold 10 slow counts. Be sure to breathe as you hold the leg and do not tense the shoulders. Lower leg straight to floor and reverse legs.

(b) TWO LEG LIFT-UP
slightly more advanced

Beginning in starting position, lower legs straight to floor; slowly raise both legs upward to vertical position, hold 10 counts, slowly lower to floor.

IV. PELVIC LIFT-UP
for thighs, back, and abdomen

Beginning in starting position, slowly lift upward to maximum, tensing buttocks; hold for 10 counts, slowly lower back to floor.

V. BALL-UP POSITION
for back, legs, and abdomen

Bring knees to chest, cross ankles, clasp hands around ankles and hold 10 counts.

Program Day Two

Menu

BREAKFAST

> ½ grapefruit or 1 4-oz. glass of fresh orange juice
> ½ cup serving cereal (suggested list, page 19)
> ½ cup low-fat milk
> 1 tsp honey (optional)
> Herb tea

LUNCH

> Mushroom, watercress, and endive salad
> Individual cheese soufflés with herbs
> Frozen peach yogurt Melba
> Herb tea

DINNER

> Oriental watercress soup
> Southeast Asian-style beef kabobs served in lettuce
>   and herb wrappings with peanut sauce
> Steamed honey pears
> Herb tea

Exercises

- *SPINE ROCK*
- *OFF-BALANCE SITTING POSITION*
- *FROG BALANCE*
- *TENT POSITION*

Mushroom, Watercress, and Endive Salad

105 Calories • 3.7 gr. Carbohydrates • 10.75 mg. Sodium

INGREDIENTS

1 bunch watercress
2 endive
8 large, white, very closed mushrooms
2 tsp vinegar (wine or brown rice) or lemon juice
3 Tb safflower oil
½ tsp minced scallion (green onion)
1 tsp fresh minced parsley
Pepper and paprika to taste

METHOD

1. Swish watercress in cold water to wash; dry with paper towels. Remove tough stems.

2. Core endive and wash leaves under cold running water. Pat dry with paper towels.

3. Wipe off mushrooms with a damp paper towel. Cut off the discolored bottom of stems and slice mushrooms into thin umbrella shapes.

4. Arrange vegetables attractively on four salad plates.

5. Whisk oil into vinegar or lemon juice in a stream.

6. Spoon dressing over salads and sprinkle with herbs and seasonings.

7. Serve slightly chilled or at room temperature.

Individual Cheese Soufflés with Herbs

189 Calories • 5.3 gr. Carbohydrates • 13.8 mg. Sodium

INGREDIENTS

1 tsp unsalted butter
1 cup skim milk
2 tsp cornstarch
2 egg yolks
½ cup grated Swiss Lace (low-sodium Swiss) or Jarlsberg cheese
1 Tb fresh minced basil or ½ tsp dry sweet basil
1 Tb fresh snipped chives (cut with scissors)
2 tsp grated Parmesan cheese
Pepper and nutmeg to taste
3 egg whites

Garnish:
2 tsp grated Parmesan cheese

METHOD

1. Preheat oven to 400° and lightly butter 4 individual 1-cup soufflé dishes or 1 1-qt. soufflé dish.

2. Place milk in a heavy saucepan, *not aluminum,* and simmer for 1 minute.

3. Whisk egg yolks and cornstarch together in a 1-qt. mixing bowl until pale yellow and frothy.

4. Whisking egg mixture constantly, add hot milk in a stream.

5. Return mixture to saucepan. Place over moderate heat and stir constantly with a wooden spatula until slightly thickened, approximately 30 seconds.

6. Add cheeses and herbs, continuing to stir until mixture is thickened and creamy, approximately 1 minute.

7. Remove cheese mixture from heat and return to mixing bowl.

8. In a separate bowl, beat egg whites until they stand in stiff peaks but are not dry.

9. Add one third of whites to warm cheese mixture and fold in thoroughly with a rubber spatula. Then gently fold in remaining whites and carefully spoon mixture into soufflé dishes. Use rubber spatula to gently smooth the top of soufflés, leaving them slightly domed in the middle. Sprinkle with remaining Parmesan cheese.

10. Place soufflés in the middle of preheated oven. Reduce to 375° and bake for 8 to 10 minutes or until soufflés are slightly firm and lightly browned on top. If making one large soufflé, allow 18 to 20 minutes baking time.

11. Serve immediately.

Frozen Peach Yogurt Melba

84 Calories • 18 gr. Carbohydrates • 29 mg. Sodium

INGREDIENTS

2 peaches (1 cup purée)
1 tsp lemon juice
2 Tb honey or to taste (1 Tb = 61 calories, 16.5 gr. carbohydrates, 1 mg. sodium)
⅔ cup plain yogurt

Garnish:
½ cup fresh raspberries or sliced fresh strawberries
Fresh mint

METHOD

1. Plunge peaches into boiling water for a few seconds to loosen skin, then peel and pit.

2. Toss peach flesh with lemon juice and purée.

3. Combine peach purée with honey, then fold in yogurt.

4. Freeze yogurt mixture in an ice cream freezer, following the instructions for your machine, or put yogurt in a covered container and place in freezer.

5. If you are not using an ice cream freezer, the frozen mixture will be very firm. A fluffier texture can be obtained by processing the frozen mixture briefly in a food processor with steel blade before serving or by beating the mixture with a whisk 2 or 3 times during the freezing process.

6. Layer fresh berries and frozen yogurt in chilled stemmed wine glasses. Garnish the top of each glass with a sprig of mint. Serve at once.

PROGRAM DAY TWO

Oriental Watercress Soup

38 CALORIES • 4.13 GR. CARBOHYDRATES • 56.0 MG. SODIUM

INGREDIENTS

4½ cups low-sodium chicken broth, homemade (see recipe) or canned
1 bunch watercress, washed, large stems removed. Leaves coarsely chopped
2 scallions, thinly sliced. Reserve 1 Tb for garnish
¼ tsp grated fresh ginger root or a pinch of dry ginger
2 tsp grated lime zest (green only)
1 Tb distilled white or rice vinegar
1 Tb dry sherry
2 Tb fresh lime juice or to taste
Pepper to taste

GARNISH:
1 Tb reserved sliced scallion
4 lime wedges

METHOD

1. Bring chicken broth to a boil.

2. Add watercress and scallion. Simmer 2 minutes.

3. Add ginger and lime zest. Simmer 1 minute.

4. Stir in vinegar, sherry, and lime juice and pepper to taste.

5. Serve soup hot, garnished with sliced scallion and lime wedges.

PROGRAM DAY TWO

Southeast Asian-Style Beef Kabobs Served in Lettuce and Herb Wrappings with Peanut Sauce

259 CALORIES • 14 GR. CARBOHYDRATES • 29 MG. SODIUM

INGREDIENTS

1 lb. flank steak

MARINADE:
Zest of 1 lime finely minced
1 tsp lime juice
2 tsp minced garlic
1 tsp honey
1 Tb minced shallot or green onion (white only)
1 tsp sesame oil
3 Tb white distilled or rice vinegar
1 tsp minced fresh ginger root

PEANUT SAUCE:
6 Tb white distilled or rice vinegar
4 tsp honey
2 Tb finely chopped unsalted peanuts (for carrot sauce, substitute 2 Tb grated carrot for peanuts)

GARNISH:
1 to 2 heads of Boston or Bibb lettuce (5 to 6 leaves per person)
1 large cucumber, peeled, seeded, and thinly sliced
4 cloves of garlic, peeled and sliced paper thin
6 radishes sliced into thin rounds
1 cup bean or alfalfa sprouts
1 well-washed bunch of fresh coriander, parsley, or basil sprigs
1 well-washed bunch of mint sprigs
2 doz 8- or 10-inch bamboo skewers

METHOD

1. Slice steak against the grain into ribbons ⅛ inch thick. Place in a shallow bowl.

2. Combine marinade ingredients and pour over steak. Marinate for at least 1 hour at room temperature or, for even better flavor, 1 to 2 days refrigerated.

3. When ready to serve, soak bamboo skewers in water for a few minutes and dry with paper towels. Thread beef ribbons on to skewers.

4. Prepare sauce.

5. Arrange garnishes and wrappings of lettuce, garlic, radishes, sprouts, and herbs decoratively on a serving platter.

6. Grill beef kabobs over charcoal or in a preheated broiler for 1 to 2 minutes on each side.

7. Using a lettuce leaf as a wrapper, make packages of beef with vegetables and herbs to taste. Dip in peanut sauce.

PROGRAM DAY TWO

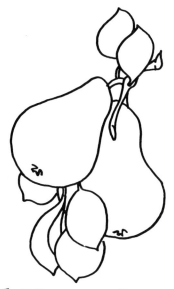

Steamed Honey Pears

140 Calories • 33.7 gr. Carbohydrates • 4.0 mg. Sodium

INGREDIENTS

1 Tb honey
1 Tb lemon juice
⅛ to ¼ tsp ground ginger to taste
4 medium-size, unblemished, fine-textured pears, not too ripe

METHOD

1. Combine honey, lemon juice, and ground ginger in a small mixing bowl.

2. Core pears from the bottom, leaving top stem intact, then peel.

3. Baste pears with honey mixture, using a pastry brush.

PROGRAM DAY TWO

Reserve a small amount of mixture to glaze pears just before serving.

4. If using a "center funnel steamer," place pears directly in steamer. If using a "basket steamer," or if improvising, place pears in a heat-proof dish large enough to hold them and catch juices.

5. Place dish in basket of basket steamer over boiling water. If improvising, use heat-proof custard cups or empty tuna cans with both ends removed to keep dish with pears above boiling water in a large saucepan with lid.

6. Cover steamer or saucepan and steam pears, basting frequently with juices, until tender when pierced with a knife — between 20 and 30 minutes, depending on ripeness of pears.

7. Before serving, brush pears with remaining honey mixture to glaze.

8. Serve pears warm, room temperature, or chilled in wine glasses or compotes, with juices poured around them.

I. SPINE ROCK
stimulates circulation, limbers spine and legs

(a) Start from the sit-up, crossed-leg position, spine relaxed, head dropped forward; clasp toes.

(b) Inhale as you roll back, bringing toes close to floor; exhale as you roll forward to starting position. Repeat 6 times.

II. OFF-BALANCE SITTING POSITION
for buttocks and thighs

(a) Sit on heels, hands resting at top of thighs.

(b) Open heels and sit with buttocks resting on floor.

(c) Lift body and sit with buttocks resting on left heel; return to center (II b); lift body to right heel, return to center (II b).

(d) OFF-BALANCE SITTING POSITION VARIATION

Start as in position II a, but with arms held upwards; bring the weight of body to the floor left; return to center (II a), then to right. Return to center (II a).

III. FROG BALANCE
tones muscles of thighs, legs, and feet

Starting from a standing position, feet together, arms stretched out in front at shoulder level, slowly lower body to squatting position, buttocks almost resting on heels, heels off floor; raise arms above head and clasp hands, hold for 10 counts.

IV. TENT POSITION

tones muscles of legs and feet. Stretches and strengthens muscles on back of thighs.

(a) Starting from position II a, place hands on floor in front of knees and lift upward on toes, hold 5 counts.

(b) Slowly squeeze heels close to floor and hold 5 counts.

Program Day Three

Menu

BREAKFAST

>Banana or peach energy drink
>Herb tea

LUNCH

>Orange and onion salad
>Puffy asparagus and cheese omelet
>Louisiana honey custard
>Herb tea

DINNER

>Chilled curried fruit soup
>Breast of chicken tandoori
>Kitcheri (brown rice cooked with lentils)
>Ananas glacé (pineapple sherbet served in a pineapple)
>Herb tea

Exercises

- SHOULDER STAND
- PELVIC LIFT-UPS
- BACK LEG LIFTS

PROGRAM DAY THREE

Banana or Peach Energy Drink

SERVES 1

193 Calories • 32.2 gr. Carbohydrates • 53.3 mg. Sodium

INGREDIENTS

1 small banana or 1 ripe fresh peach, peeled and pitted
1 tsp lemon juice
4 Tb plain yogurt
2 tsp toasted wheat germ
1 tsp honey (optional)
1 tsp vanilla-flavored instant protein powder (optional)
2 ice cubes

METHOD

1. Combine above ingredients in a blender or a food processor with steel knife.

2. Serve as a quick-nourishing breakfast or lunch.

Orange and Onion Salad

SERVES 4

92 CALORIES • 15.7 GR. CARBOHYDRATES • 1.75 MG. SODIUM

INGREDIENTS

3 navel oranges
8 thin slices of red onion divided into rings
1 Tb olive oil
1 tsp lemon juice
Freshly ground black pepper
4 lettuce leaves (Bibb or Boston)

METHOD

1. Using a sharp knife, carefully peel oranges, removing white inner skin. Divide orange into skinless sections.

2. Place orange sections and onion rings in a bowl.

3. Sprinkle with olive oil, lemon juice, and black pepper to taste.

4. Allow salad to marinate for at least ½ hour before serving.

5. Place a lettuce leaf on each salad plate and arrange orange sections in a fan shape on lettuce. Top oranges with onions and spoon over any juice.

PROGRAM DAY THREE

Puffy Asparagus and Cheese Omelet

112 C ALORIES • 13.5 GR . C ARBOHYDRATES • 226.4 MG . S ODIUM

INGREDIENTS

4 large eggs, separated
16 spears fresh asparagus
2 Tb grated Parmesan cheese
1 Tb minced shallot or onion
1 Tb minced parsley
Freshly ground pepper to taste
1 6- to 8-inch non-stick omelet pan with a metal handle or
1 6- to 8-inch well-seasoned iron omelet pan and 2 tsp safflower oil

METHOD

1. Cut woody ends off asparagus. Steam spears lightly, not more than 1 minute. Plunge into cold water for a few seconds to preserve their color. Then dry with paper towels.

2. Allow asparagus to cool, cut off 3-inch tips and reserve. Chop the rest of the stalks into small pieces.

3. Beat egg yolks until light and frothy. Stir in chopped asparagus, cheese, shallot, parsley, and pepper.

4. Beat egg whites until they stand in soft peaks, then carefully but thoroughly fold whites into yolk mixture.

5. Preheat oven to 300°.

6. If using a non-stick pan, heat it slightly, then spoon in ¼ of omelet mixture. If using an iron pan, cover with a light film of oil, then heat and spoon in mixture.

7. Cook omelet over medium heat until bottom is lightly browned. Make sure to run a small knife around outside of omelet to loosen edges.

8. Arrange 4 reserved asparagus tips on ½ of omelet top.

9. Place omelet in preheated oven until top is just set, approximately 1 minute.

10. Fold omelet in half and turn onto a heated plate. Keep warm.

11. Repeat steps 6–10 until all of mixture is used.

12. Serve at once.

PROGRAM DAY THREE

Louisiana Honey Custard

94 CALORIES • 11.8 GR. CARBOHYDRATES • 49.5 MG. SODIUM

INGREDIENTS

1 egg and 1 yolk
2 Tb honey
⅛ tsp vanilla
1 cup low-fat milk
Dash of nutmeg and mace or cinnamon (optional)

METHOD

1. Preheat oven to 375°.

2. Place egg, yolk, honey, and vanilla in a 1-qt. mixing bowl and whisk until yellow and frothy.

3. Heat milk in a small saucepan to just below a boil.

4. Whisking egg mixture constantly, add hot milk in a stream.

5. Pour custard mixture through a wire mesh strainer into 4 ovenproof ramekins or custard cups.

6. Skim any foam off the top of custard with a small spoon to assure a smooth surface, then sprinkle lightly with nutmeg and mace or cinnamon.

7. Place ramekins in an ovenproof container with sides at least 2½" high. Fill container with enough boiling water to reach halfway up the sides of ramekins.

8. Place container on the middle rack of a preheated oven. Reduce heat to 350° and bake for 30 minutes, or until custard is set.

9. Serve custard chilled or at room temperature.

PROGRAM DAY THREE

Chilled Curried Fruit Soup

127 Calories • 29 gr. Carbohydrates • 21.6 mg. Sodium

INGREDIENTS

2 ripe mangoes, peeled and pitted
1 ripe peach, peeled and pitted
2 Tb lemon juice
½ cup orange juice freshly squeezed
½ tsp safflower oil
2 tsp minced onion
1 tsp curry powder
⅛ tsp powdered ginger
1 cup low-sodium chicken broth and 1 cup water, or
  2 cups water
Dash of cinnamon
Pepper to taste

Garnish:
½ unpeeled diced red Delicious or other firm red
  eating apple
1 tsp minced parsley
1 tsp snipped chives

METHOD

1. Toss mango and peach flesh with lemon juice and purée in a food processor, blender, or foodmill.

2. Pour purée into a large mixing bowl and add orange juice.

3. Lightly coat a non-stick skillet with ½ tsp safflower oil and add minced onion, curry powder, and ginger.

4. Cook onion mixture over low heat, stirring frequently until onion is soft but not browned — approximately 1 minute.

5. Add 1 cup chicken broth or water to skillet and simmer for 1 minute, scraping the bottom of skillet with a wooden spatula to make use of all of onion mixture.

6. Add mixture in skillet and 1 cup cold water to fruit purée.

7. Season with cinnamon and pepper to taste. Chill.

8. Serve chilled soup garnished with diced apple, parsley, and chives.

PROGRAM DAY THREE

Breast of Chicken Tandoori

218 Calories • 4.65 gr. Carbohydrates • 98.3 mg. Sodium

INGREDIENTS

4 half-breasts of chicken, approximately ⅓ lb. each weighed with bone

Marinade:
1 Tb cider vinegar
2 Tb safflower oil
1 Tb minced dried apricot
¼ cup plain yogurt
1 tsp caraway seeds
1 tsp ground coriander
1 tsp minced garlic
2 tsp minced fresh ginger or ½ tsp powdered ginger
½ tsp ground cumin
1 Tb fresh lime or lemon juice

Garnish:
Dash of paprika
4 lemon or lime wedges
4 sprigs fresh coriander (Chinese parsley) or broad-leaf Italian parsley

METHOD

1. Skin chicken breasts and remove any fat.

2. With a small sharp knife, make an incision in the side of each half breast, forming a pocket.

3. Place chicken in a shallow ovenproof dish large enough to hold it in 1 layer.

4. Combine marinade ingredients in a small bowl. Pour over chicken, making sure marinade covers chicken and is in the incisions.

5. Marinate chicken overnight, refrigerated, or for at least 8 hours at cool room temperature.

6. Preheat oven to 400°.

7. Place chicken in its original dish in the middle of the preheated oven, meat side up, lightly sprinkled with paprika.

8. Bake for 25 to 30 minutes.

9. Serve at once garnished with fresh coriander and lime wedges, accompanied by either kitcheri or brown rice pilaf.

Kitcheri (Brown Rice Cooked with Lentils)

176 CALORIES • 30 GR. CARBOHYDRATES • 9.45 MG. SODIUM

INGREDIENTS

½ cup orange Egyptian lentils (available in Indian, Middle Eastern, and health food stores)
½ cup long grain brown rice
2 tsp safflower oil
⅔ cup finely chopped onion
1½ tsp finely minced fresh ginger root
1 tsp finely minced garlic
½ tsp turmeric
1¼ to 1½ cups low-sodium chicken broth or water
2 Tb plain yogurt (optional)
1½ tsp fresh snipped chives or scallion greens

METHOD

1. Pick over lentils and rice and remove any impurities such as pebbles or discolored grains.

2. Wash lentils and rice together in cold running water until water runs clear.

3. Cover with cold water and soak for 1 hour.

4. Place oil in a non-stick skillet and add onion, ginger, and garlic. Cook over medium heat, stirring often until onion is golden.

5. Thoroughly drain rice and lentils and add them to the mixture in the skillet.

6. Stirring constantly, cook for 2 minutes, then add turmeric and cook for 1 minute more. Meanwhile bring broth to a boil in small saucepan.

7. Place mixture in a heavy saucepan and add boiling broth or water to cover rice and lentils by ½ inch.

8. Place over medium heat and when liquid returns to a vigorous boil cover and reduce heat to low.

9. Cook for 15 to 20 minutes or until liquid is absorbed and rice and lentils are tender. Add more liquid if they are not tender and cook a bit more.

10. Serve portions of kitcheri topped with a dollop of yogurt (optional) and snipped chives.

PROGRAM DAY THREE

Ananas Glacé (Pineapple Sherbet Served in a Pineapple)

SERVES 6
64 CALORIES • 16.05 GR. CARBOHYDRATES • 17 MG. SODIUM

INGREDIENTS

1 large ripe unblemished pineapple with an attractive top or 1 20-oz. can crushed pineapple in unsweetened juice
Juice of 2 limes — 6 Tb
2 Tb honey
¾ cup cold water or pineapple juice
1 egg white

GARNISH:
Fresh mint

METHOD

1. Cut off the tufted top of the pineapple ¾ of an inch below the crown and reserve it as a lid.

2. Very carefully hollow out the pineapple with a sharp knife and spoon, being sure not to break the skin. Refrigerate the shell.

3. Remove core and chop remaining flesh into small cubes. Reserve ⅓ cup and purée the rest along with ½ cup juice. You should have approximately 2 cups of purée.

4. Stir in lime juice, honey, and either ⅔ cup water or juice.

5. Freeze mixture in an ice cream freezer, following your machine instructions. Halfway through freezing process add ½ egg white that has been beaten until it stands in soft peaks. Continue freezing until texture is firm but fluffy. Mix with reserved pineapple chunks and fill pineapple shell.

6. If you do not have an ice cream freezer, sherbet can also be made in ice trays that have been thoroughly washed, then rinsed with cold water and baking soda to remove all odors.

7. If using the "ice tray" method, fill trays only ⅔ full and beat sherbet with a whisk or fork 2 or 3 times during freezing to make its texture fluffier. Halfway through freezing, whisk in ½ of the beaten egg white.

8. Whichever method you employ, if sherbet is too firm to spoon easily into pineapple shell process briefly in a food processor with steel blade before mixing with the pineapple chunks.

9. For an attractive presentation fill pineapple shell with sherbet and replace lid.

10. Spoon sherbet into chilled wine glasses and garnish with fresh mint.

I. SHOULDER STAND

reverses the stress caused by the downward pull of gravity. An important position for improving circulation in the legs.

(a) Start from ball-up position (knees close to chest), roll back and up into shoulder stand.

(b) Hands supporting back, hold 20 counts.

(c) Drop knees to forehead, arms at side; slowly roll back and legs down to floor.

II. PELVIC LIFT-UPS
for firm thighs and buttocks

(a) Lift buttocks as high as possible, arms straight overhead, hold 10 counts, return to floor.

(b) PELVIC LIFT-UP VARIATION
slightly more advanced

Begin as in IIa but lift heels from floor and rest weight on toes for even more stretch.

III. BACK LEG LIFTS
for buttocks and thighs

(a) Fold arms on floor, head resting on wrists. Swing straight leg up and down in 10 continuous movements; rest; reverse legs.

(b) Back Leg Lift Variation

From all fours, arms held straight throughout, swing left leg up, head turned up; then draw the left knee to chest and forehead to knee. Return to starting position and repeat 4 times; reverse legs.

Program Day Four

Menu

BREAKFAST

> ½ grapefruit
> 2 oz. cheese (suggested list page 18)
> 1 slice gluten toast (optional)
> 1 tsp whipped unsalted butter (optional)
> 1 tsp honey (optional)
> Herb tea

LUNCH

> Yucatecan chicken and lime soup
> Strawberries in orange baskets
> Herb tea

DINNER

> Cold cucumber-yogurt soup
> Baked fish steaks Moroccan style
> Papaya boats
> Herb tea

Exercises

- SIDE LEG LIFTS
- BACK LEG LIFTS
- KNEE PRESS-DOWNS
- FORWARD STRETCH

PROGRAM DAY FOUR

Yucatecan Chicken and Lime Soup

116 Calories • 8.0 gr. Carbohydrates • 127.0 mg. Sodium

INGREDIENTS

6 cloves garlic
1 tsp fresh minced oregano or ¼ tsp dried
6½ cups homemade chicken broth (see recipe, p. 22) or low-sodium canned
¼ tsp ground cumin
1 chicken breast, approximately ½ lb.
½ tsp safflower oil
⅓ cup finely chopped onion
¼ cup finely chopped green pepper
1 large ripe tomato, approximately ½ lb., skinned, seeded, and chopped
1 tsp white vinegar
1 tsp lime zest
1 Tb fresh lime juice
Freshly ground pepper to taste

Garnish:
4 lime wedges
1 tsp minced parsley

METHOD

1. Toast unpeeled garlic cloves by placing in a preheated, heavy, non-stick skillet over medium heat. Brown both sides,

approximately 1 minute each side. Remove from skillet and put aside until cool enough to peel.

2. After garlic is removed, place oregano in skillet over medium heat for approximately 30 seconds, stirring from time to time.

3. Add peeled garlic and oregano to chicken broth and bring to a boil in a large saucepan with cover.

4. Add cumin and chicken breast, and simmer partially covered for 15 minutes or until chicken is tender.

5. Strain the broth and put chicken breast aside until cool enough to handle. Skin breast and remove meat. Shred meat and reserve.

6. Heat oil in the non-stick skillet. Add onion and pepper, and sauté until soft but not browned.

7. Add the tomato to skillet and continue to cook for five minutes more.

8. Return broth to saucepan. Add tomato mixture and simmer for 5 minutes.

9. Stir in shredded chicken, vinegar, and lime zest, and simmer 2 to 3 minutes more. Add lime juice and pepper to taste.

10. Serve soup hot, garnished with lime wedges and chopped parsley.

Strawberries in Orange Baskets

80 Calories • 19.7 gr. Carbohydrates • 0 mg. Sodium

INGREDIENTS

4 unblemished oranges
2 cups fresh strawberries
1 Tb honey

Garnish:
Fresh mint

METHOD

1. Using a small, sharp knife to cut and a grapefruit spoon

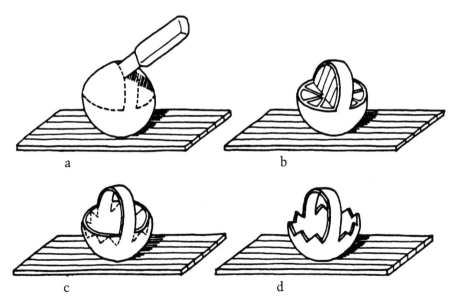

a

b

c

d

to hollow, make orange baskets, following steps in the diagrams.

2. Wash strawberries and cut in half.

3. Squeeze orange pulp to remove juice. Mix juice with honey and add to strawberries. Refrigerate for at least ½ hour.

4. Fill baskets with chilled berries and orange juice mixture. Serve baskets on beds of fresh mint.

PROGRAM DAY FOUR

Cold Cucumber-Yogurt Soup

117 CALORIES • 14.0 GR. CARBOHYDRATES • 67.0 MG. SODIUM

INGREDIENTS

1 European-style seedless cucumber or 2 medium regular cucumbers
1 clove garlic
1 Tb wine vinegar
1 Tb good quality olive or safflower oil (preferably cold pressed)
1 pt. plain low-fat yogurt
2 Tb minced fresh mint or 1 tsp dried
2 Tb minced fresh parsley
2 Tb minced green onion, green and white parts
1 Tb currants or chopped raisins
Pepper to taste
½ cup cold water or 6 ice cubes

GARNISH:
6 medium-size radishes
1 Tb currants or chopped raisins
4 sprigs fresh mint or parsley

METHOD

1. Do not peel cucumber unless it has been waxed. Slice into rounds ⅛" thick, then cut rounds into thin strips. You should have 2 cups of strips.

2. Cut garlic clove in half and rub it around the inside of a 1-qt. mixing bowl.

3. Add vinegar and oil and swish around to flavor slightly with garlic.

4. Stir in yogurt, blending well, then add cucumber strips, mint, parsley, onion, currants, and pepper to taste.

5. If soup need not be served immediately, stir in cold water and chill for at least one hour. If pressed for time, crush ice cubes and add to soup to thin and chill it.

6. While soup is chilling, slice radishes into paper-thin rounds and stir into soup just before serving, reserving a few for garnish.

7. Serve chilled soup garnished with radish rounds, currants, and sprigs of mint.

PROGRAM DAY FOUR

Baked Fish Steaks Moroccan Style

262 Calories • 8.39 gr. Carbohydrates • 165 mg. Sodium

INGREDIENTS

1½ lb. halibut, sea bass, or red snapper steaks, cut approximately ¾" thick

Marinade:
¼ cup finely minced fresh coriander (cilantro) and ¼ cup finely chopped parsley, or ½ cup finely chopped parsley
1 Tb white distilled vinegar
2 to 3 cloves garlic, peeled and very finely chopped
1½ tsp ground cumin
2 tsp paprika
pinch of cayenne
2 Tb safflower oil
¼ cup lemon juice

6 ribs of celery, washed and leaves removed
2 ripe but firm tomatoes, unpeeled
1 Tb minced fresh parsley
1 Tb minced fresh mint (optional)
1 tsp grated lemon zest

METHOD

1. Rinse the fish under cold running water, pat dry with paper towels. Place in a shallow baking dish large enough to hold it in one layer.

2. Combine marinade ingredients in a blender or food processor and liquefy as much as possible. Pour over fish and marinate for at least ½ hour.

3. While fish marinates, remove any strings from celery ribs with a vegetable peeler or small sharp knife. If ribs are wide, cut in half lengthwise. Then cut into pieces 2 to 3 inches long.

4. Slice tomatoes into rounds ¼" thick.

5. Preheat oven to 400°.

6. Arrange celery in one layer in the bottom of a lightly oiled baking dish. Arrange the fish steaks on top of celery and top with a layer of tomato slices.

7. Pour marinade over fish and vegetables and sprinkle with ½ of reserved chopped parsley, mint, and lemon zest. Reserve the other ½ to garnish the completed dish.

8. Cover with aluminum foil and bake in the middle of preheated oven for 30 minutes.

9. Remove the foil and spoon marinade over fish and tomatoes to moisten.

10. Turn oven heat to 550° and place fish uncovered on the top shelf of oven for 15 to 20 minutes or until fish is flaky and top of tomatoes nicely browned.

11. Serve at once garnished with reserved parsley, mint, and lemon zest.

PROGRAM DAY FOUR

Papaya Boats

30 Calories • 7.5 gr. Carbohydrates • 2.25 mg. Sodium

INGREDIENTS

1 large ripe papaya. If papaya is not available, a small melon may be substituted.

Garnish:
4 wedges lemon or lime
4 sprigs fresh mint

METHOD

1. Cut papaya in quarters lengthwise. Seed.

2. Using a sharp knife, carefully cut between the rind and

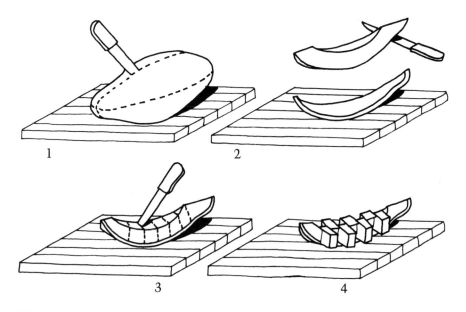

the flesh of the papaya, removing the flesh in one piece and leaving a rind ¼ inch thick.

3. Place flesh back on the rind and make horizontal cuts across the flesh to form bite-size pieces.

4. Place papaya quarters on plates and carefully push cut pieces of papaya in alternating directions so that the end result looks something like a boat with oars.

5. Garnish with lemon or lime wedges and fresh mint.

I. SIDE LEG LIFTS
for outer thigh

Lying on right side, right arm overhead, left hand on floor for balance.

(a) Lift the left leg as high as possible (keeping knee straight), lower to floor, repeat 10 times. Reverse sides by rolling over to left.

PROGRAM DAY FOUR

(b) Lift both legs up simultaneously, hold 4 counts; repeat; reverse sides.

PROGRAM DAY FOUR

II. BACK LEG LIFTS
for buttocks, lower back, and front of thighs

Lying face down, arms at side, raise left leg upward (keeping it straight); hold 4 counts, lower to floor, and reverse sides.

PROGRAM DAY FOUR

III. KNEE PRESS-DOWNS
for inner thighs and pelvic flexibility

Place soles of feet together, clasping ankles with hands; slowly press knees downward toward the floor.

IV. FORWARD STRETCH
for legs, back, and abdomen

(a) Start from sit-up position, right leg folded back, left extended.

(b) Inhale, stretching arms upward; exhale, bending forward; bring head as close to knee as possible while clasping hands around arch of foot or ankle; hold 10 counts, release, and reverse sides.

(c) Forward Stretch Variation

With both legs extended, start with arms overhead. Stretch upward, inhaling, and bend slowly forward, exhaling. Take hold of feet, forehead close to knees, hold 10 counts.

Program Day Five

Menu

BREAKFAST

> ¼ papaya or 1 slice fresh pineapple
> Huevos rancheros (baked eggs Mexican style)
> 1 slice gluten toast (optional)
> ½ tsp whipped unsalted butter (optional)
> 1 tsp honey (optional)
> Herb tea

LUNCH

> Exotic fruit salad with vinaigrette sauce
> Banana sherbet
> Herb tea

DINNER

> Green bean and onion salad
> Chicken breasts with sweet and sour sauce, bouquet of vegetables, and tart apple purée
> Lemon soufflé
> Herb tea

Exercises

- COBRA POSITION
- BOW POSTURE
- V POSTURE
- CAT HUMP RESISTANCE STRETCH
- CAMEL POSTURE
- PLOW POSTURE

PROGRAM DAY FIVE

Huevos Rancheros (Baked Eggs Mexican Style)

124 Calories • 4.82 gr. Carbohydrates • 3.17 mg. Sodium

INGREDIENTS

2 tsp safflower oil
2 Tb minced onion
2 cups diced tomato (2 medium)
⅓ cup finely diced green pepper
⅛ tsp honey
⅛ tsp ground cumin or to taste
pepper, paprika, and cinnamon to taste
4 eggs
boiling water

Garnish:
1 tsp minced fresh parsley or cilantro (coriander)

METHOD

1. Preheat oven to 400°.

2. Lightly oil 4 individual ovenproof egg baking dishes.

3. Spread remaining oil evenly over the surface of a medium-size non-stick skillet.

4. Place onion in skillet and cook over medium to low heat, stirring frequently until soft but not brown.

5. Add diced tomato and green pepper to skillet and simmer, stirring and pressing on tomato until it is nearly a purée.

6. Stir in honey, cumin, black pepper, paprika, and a pinch of cinnamon to taste.

7. Spoon ¼ of tomato sauce around the outer rim of each baking dish and break an egg into the center. Sprinkle eggs lightly with pepper and paprika.

8. Place baking dishes in a bain-marie (an ovenproof container large enough to hold them) and pour in enough boiling water to reach halfway up the sides of dishes. Place bain-marie in the middle of a preheated oven.

9. Bake eggs for 10 minutes or until whites are barely set and yolks are still soft.

10. Garnish eggs with minced parsley or cilantro and serve at once.

PROGRAM DAY FIVE

Exotic Fruit Salad with Vinaigrette Sauce

300 CALORIES • 18 GR. CARBOHYDRATES • 23.2 MG. SODIUM

INGREDIENTS

1 ripe papaya
1 large ripe avocado
½ lemon
2 ripe kiwi fruit (optional)
8 large lettuce leaves, washed and dried

VINAIGRETTE SAUCE:
1 tsp Dijon mustard
1 Tb wine vinegar
5 Tb safflower oil
1 tsp minced shallot or scallion (green onion)
½ tsp chopped fresh mint

METHOD

1. Cut papaya and avocado in half and then in quarters lengthwise. Seed.

2. Cutting through the skin with a sharp knife, divide quarters into long strips of equal width. Peel strips and squeeze lemon juice over them to prevent discoloration.

3. Peel kiwi fruit and slice into thin rounds, then into half moon shapes.

4. Place lettuce leaves on 4 plates. Arrange alternating strips of papaya and avocado decoratively on beds of lettuce and surround with slices of kiwi fruit.

5. Combine mustard and vinegar in a small bowl. Whisking constantly with a metal whisk or fork, add oil in a slow stream. The sauce should thicken slightly. Stir in minced shallot and mint.

6. Spoon vinaigrette over salads and serve chilled or at room temperature.

PROGRAM DAY FIVE

Banana Sherbet

SERVES 6
63 Calories • 17.1 gr. Carbohydrates • 2 mg. Sodium

INGREDIENTS

2 medium bananas (approximately 1 cup purée)
Juice of 1 lemon
3 Tb honey
1 cup cold water

Garnish:
1 small bunch fresh mint

METHOD

1. Place bananas, lemon juice, and honey in a blender or food processor. Blend until smooth.

2. Gradually add water while blending until mixture is light and fluffy.

3. Freeze mixture in an ice-cream freezer, following the instructions for your machine, or pour into ice trays that have been thoroughly washed with boiling water and rinsed with cold water and baking soda to remove all odors.

4. If using the ice-tray method, fill trays only ⅔ full and stir the sherbet with a whisk or fork 2 or 3 times during freezing to make it fluffier in texture. For an even fluffier texture, process sherbet briefly in a food processor with steel knife attachment just before serving.

5. Spoon sherbet into stemmed wine glasses and garnish with fresh mint.

PROGRAM DAY FIVE

Green Bean and Onion Salad

78 Calories • 6.3 gr. Carbohydrates • 6.5 mg. Sodium

INGREDIENTS

⅔ lb. fresh green beans
1 red bell pepper, long, round, and narrow in form if possible
1 large or 2 small green onions (scallions)
1½ Tb olive oil
2 tsp wine vinegar
Pepper to taste
4 attractive lettuce leaves — romaine, Boston, or Bibb

METHOD

1. Wash all vegetables under cold running water.

2. Remove strings from beans.

3. Steam beans over boiling water until just tender, approximately 1 minute.

4. Place beans under cold running water to preserve their fresh color.

5. Core bell pepper and cut into rings ⅛ inch wide.

6. Slice green onions into thin rings. You should have about 2 Tb.

7. Place well-drained beans in a bowl and add onion, olive oil, and vinegar. Toss together and sprinkle with freshly ground pepper to taste.

8. Marinate beans for at least 15 minutes.

9. When ready to assemble, remove beans from the bowl and form 4 bundles.

10. Slip a pepper ring around the center of each bundle and place bundles on chilled salad plates covered with lettuce leaves.

11. Sprinkle onion and any remaining dressing over bundles and serve.

PROGRAM DAY FIVE

Chicken Breasts with Sweet and Sour Sauce, Bouquet of Vegetables, and Tart Apple Purée

323 CALORIES • 43.0 GR. CARBOHYDRATES • 41.0 MG. SODIUM

INGREDIENTS

4 cups unsweetened apple cider
2 leeks, white and tender green only, cut into rounds ¼" thick
12 small red cabbage leaves
1 garlic clove peeled but whole
2 whole chicken breasts, approximately ¾ lb. each, unboned
4 medium carrots, peeled and cut into 2-inch lengths, then into quarters lengthwise
2 medium zucchini, unpeeled, cut into rounds ½ inch thick

APPLE PURÉE:
2 tart cooking apples peeled, cored, and quartered
1 cup apple cider
Zest of one lemon

SWEET AND SOUR SAUCE:
1 Tb red wine vinegar
1 Tb Dijon mustard

GARNISH:
Parsley

METHOD

1. Place 4 cups cider in the bottom of a basket steamer or in a large saucepan with a cover, into which you can fit a colander or strainer for steaming. Bring cider to a boil.

2. Add leek, cabbage leaves, and garlic to cider and reduce heat to a simmer.

3. After 5 minutes, add chicken and return poaching liquid to a boil. Then cover, reduce heat, and simmer for 15 minutes.

4. Place carrots in steamer basket or colander and steam over chicken poaching broth for the last 6 minutes the chicken is cooking. Three minutes before the chicken is done, add zucchini to the basket and steam lightly. Vegetables should remain crisp.

5. While chicken is poaching, place 1 cup cider, apples, and lemon zest in a small, heavy-bottomed saucepan and cook over medium heat for 8 to 10 minutes or until apple is tender.

6. Purée apple mixture and set aside. Reserve unwashed saucepan.

7. Spread ¼ of the apple purée on each plate.

8. Remove vegetables and chicken from poaching liquid and steaming basket. Discard garlic and place cabbage, carrots, zucchini, and leek in a decorative border on each plate. Keep warm.

9. When chicken is cool enough to handle, remove skin and, using a small sharp knife and your fingers, remove each ½

chicken breast from bone in one piece. Trim off any gristle and place a piece of chicken in the center of each plate.

10. Pour poaching liquid into the small, heavy reserved saucepan. Allow liquid to boil over high heat until it has reduced to ¼ the original amount.

11. Whisk in vinegar and mustard and allow the sauce to cook until it becomes syrupy and translucent.

12. Spoon sauce over chicken. Garnish plates with parsley and serve at once.

Lemon Soufflé

101 CALORIES • 16.8 GR. CARBOHYDRATES • 39.0 MG. SODIUM

INGREDIENTS

½ tsp unsalted butter
1 egg yolk
3 Tb honey (pale golden color is best in this recipe)
2 tsp cornstarch
½ cup skim milk
1¼ tsp lemon zest, finely minced
5 Tb lemon juice
3 egg whites

METHOD

1. Preheat oven to 400° and lightly butter a 3-cup soufflé dish or 4 small individual soufflé dishes.

2. Place egg yolk in a 1-quart mixing bowl and whisk until pale yellow and frothy, then add honey and cornstarch. Continue to whisk until well mixed.

3. Place milk and zest in a small, heavy saucepan, *not aluminum,* and simmer for 1 minute.

4. Whisking egg mixture constantly, add hot milk in a stream.

5. Return mixture to saucepan. Place over moderate heat and stir constantly with a wooden spatula until slightly thickened, approximately 30 seconds.

6. Add lemon juice and, continuing to stir, simmer until

mixture is thickened and creamy, approximately 30 seconds more.

7. Remove lemon mixture from heat and scrape into a medium-size mixing bowl.

8. In a separate bowl, beat egg whites until they stand in stiff peaks but are not dry.

9. Add half of whites to warm lemon mixture and mix thoroughly with a rubber spatula. Fold in remaining whites gently, and carefully spoon mixture into soufflé dish or dishes. Using rubber spatula, smooth off top, leaving it slightly domed in the middle.

10. Place soufflé in the middle of preheated oven. Reduce heat to 375° and bake for 12 to 14 minutes or until soufflé is slightly firm and lightly browned on top. If making small soufflés, allow between 8 and 10 minutes.

11. Serve immediately.

I. COBRA POSITION
stretches and strengthens spine and back muscles

Start from the lying-down position, hands close to your shoulders; slowly lift upward, arching as far back as possible, head turned upward, thighs on floor; hold 10 counts, slowly lower to floor.

II. BOW POSTURE

Releases tension in the shoulders, reduces fat around the abdomen and hips, and slims the thighs.

Lie on your stomach, face down; grasp ankles, inhaling; slowly lift the upper torso and legs off the floor; hold six counts. Exhale, slowly lowering to the floor.

III. V POSTURE
for flexibility of the hip joint and toning up outer thigh

(a) From sitting position, bring left leg over right, squeeze legs together, and keep feet close to buttocks.

(b) V POSTURE VARIATION clasp hands, stretching arms upward.

IV. CAT HUMP RESISTANCE STRETCH
stretches the entire spine and strengthens abdominal muscles

(a) Start from all fours position, spine relaxed, head up.

(b) Exhale, lifting up as high as possible, head down, abdomen contracted; keep arms straight throughout; hold 10 counts. Slowly release to starting position, letting spine relax and lifting head up.

V. CAMEL POSTURE
strengthens the thigh and abdominal muscles

Start fron the kneel-sitting position, arms at your side, knees apart; place hands on heels, gently lift up, pushing the hips forward, arching the spine, letting the head drop back. Hold 6 counts and return to starting position.

VI. PLOW POSTURE
stretches and tones muscles of legs and back

(a) Starting from shoulder stand, bring both legs back to floor, arms at your side; hold 10 counts.

(b) PLOW POSTURE VARIATION
slightly more advanced

With both legs brought back to floor, bring arms overhead, hands touching toes; hold 10 counts; leaving arms overhead, slowly roll the back down to floor, extend legs, rest.

Program Day Six

Menu

BREAKFAST

> ½ grapefruit
> 2 oz. low-sodium cottage cheese (other pot cheeses listed page 18)
> 1 slice gluten toast (optional)
> ½ tsp whipped unsalted butter (optional)
> 1 tsp honey (optional)
> Herb tea

LUNCH

> 4-oz. glass fresh or low-sodium canned vegetable juice
> Eggs parmentière (eggs baked in potatoes) with herbed yogurt hollandaise sauce
> Melon swan filled with fruit
> Herb tea

DINNER

> Cold asparagus with vinaigrette sauce
> Lamb chops Provençal
> Tomatoes Provençal
> Cold strawberry yogurt soufflé
> Herb tea

Exercises

- *LOTUS POSTURES*
- *STANDING STRETCHES*

Eggs Parmentière (Eggs Baked in Potatoes) with Herbed Yogurt Hollandaise Sauce

214 Calories • 18 gr. Carbohydrates • 78.7 mg. Sodium

INGREDIENTS

2 evenly shaped Idaho baking potatoes
4 eggs (medium size is best for this recipe)

Herbed Yogurt Hollandaise Sauce:
2 medium egg yolks
1 tsp cornstarch
¾ cup plain yogurt

PROGRAM DAY SIX

HERBED YOGURT HOLLANDAISE SAUCE: *(continued)*
1 tsp fresh basil, chervil, tarragon or dill finely minced. If fresh herbs are not available, ¼ tsp of dried sweet basil, chervil, tarragon or dill may be substituted
1 tsp fresh minced parsley
1 tsp fresh snipped chives or 1 tsp finely minced scallion (green part only)
1 tsp sweet butter
Black pepper and a few drops of lemon juice to taste
1 tsp safflower oil to oil baking dish

GARNISH:
Black pepper and paprika
4 sprigs parsley

METHOD

1. Preheat oven to 425°.
2. Scrub potatoes with a vegetable brush under cold running

water. Dry with paper towels and place in the middle of the preheated oven.

3. Bake potatoes for 45 minutes or until slightly soft when squeezed. Halfway through baking time, pierce potatoes once with a fork to allow steam to escape and prevent the potato from bursting.

4. While potatoes are baking, whisk together yolks, cornstarch, and yogurt and place in the top of a double boiler.

5. Fill the bottom of double boiler with water and bring to a simmer.

6. Cook the sauce over simmering water, stirring constantly with a wooden spatula until it begins to thicken (approximately 1 minute).

7. Add the herbs and continue to cook sauce, stirring constantly 1 minute more or until creamy and medium thick.

8. Whisk in butter, pepper, and lemon juice to taste. Sauce may be kept warm over warm (not hot) water for up to 1 hour, whisking occasionally.

9. When potatoes are done, remove from oven and cut in half lengthwise with a sharp knife.

10. Scoop out potato halves with a spoon, leaving a thin shell. Rice or mash potato and mix with 2 to 3 Tb of sauce.

11. Season potato with pepper to taste and return to shells. Place potato shells in an oiled baking dish.

12. Make an indentation in the middle of each filled shell with the back of a large spoon. Break an egg into each indentation and sprinkle egg lightly with pepper if desired.

13. Bake eggs, loosely covered with foil, in a preheated 425° oven for 20 minutes or until yolk is soft but white is set.

14. Spoon sauce over eggs, sprinkle lightly with paprika and garnish with parsley sprigs. Serve at once.

PROGRAM DAY SIX

Melon Swan Filled with Fruit

90 CALORIES • 15 GR. CARBOHYDRATES • 17 MG. SODIUM

INGREDIENTS

1 large ripe flavorful honeydew, canary, Persian, or cantaloupe melon
½ cup grapes
½ cup strawberries
1 tsp lime juice

GARNISH:
Fresh mint

METHOD

1. Following the diagram below, draw a swan pattern on the melon with a pencil.

a

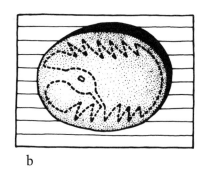

b

c   d

2. Using a small sharp knife, cut around pattern, making sure to leave the tip of the beak attached to the top of the wing.

3. Remove cut sections, cutting them into smaller pieces if necessary for easier removal.

4. Discard seeds and hollow out swan with a melon-ball cutter.

5. Toss melon balls with grapes, strawberries, and lime juice, and refill the swan with fruits. Garnish with mint.

6. Chill until ready to serve.

Cold Asparagus with Vinaigrette Sauce

181 Calories • 25 gr. Carbohydrates • 10.3 mg. Sodium

INGREDIENTS

1 lb. fresh asparagus of uniform thickness (choose stalks that are firm and crisp, with tightly closed tips)

Vinaigrette Sauce:
1 tsp Dijon mustard
1 Tb wine vinegar
5 Tb safflower oil
1 tsp minced shallot or scallion (green onion)

Garnish:
4 to 8 lettuce leaves, washed and dried
1 red sweet bell pepper, washed and cored
1 Tb minced parsley
Pepper and paprika to taste

METHOD

1. Wash asparagus under cold running water. Cut off 1" at the butt end of stalks.

2. If stalks are not the most slender and tender ones, peel them to within 3 inches of the tip with a vegetable peeler or small sharp knife.

3. Lightly steam asparagus over boiling water in an aspara-

gus or vegetable steamer, or in a colander that fits into a large pot with a lid. Cooking should take between 1 and 5 minutes, depending on the thickness of stalks.

4. Be careful not to overcook. Asparagus should remain slightly crisp. Plunge them into cold water for a few seconds to preserve their color, then drain on paper towels.

5. Combine mustard and vinegar in a small bowl. Whisking constantly with a metal whisk or fork, add oil in a slow stream. Vinaigrette should thicken slightly. Stir in shallot.

6. Cut bell pepper into rings ¼" thick. Cut pepper rings in half.

7. Place lettuce on 4 plates. Arrange asparagus on lettuce and garnish with bell pepper. Spoon over vinaigrette and sprinkle with parsley, pepper, and paprika.

PROGRAM DAY SIX

Lamb Chops Provençal

250 CALORIES • 3 GR. CARBOHYDRATES • 80.0 MG. SODIUM

INGREDIENTS

8 lean rib chops, approximately 5 oz. each, weighed with bone
1½ tsp minced garlic
1 tsp dried mint
2 tsp dried thyme
2 Tb fresh minced parsley
4 tsp lemon juice
2 tsp safflower or olive oil
Fresh ground pepper to taste

GARNISH:
4 small bouquets watercress
4 lemon wedges

Kitchen string

METHOD

1. Preheat oven to 425°.

2. Trim as much fat as possible off chops.

3. Combine garlic, mint, thyme, parsley, lemon juice, and oil, and spread on both sides of chops.

4. Press chops together as they would be in a rack of lamb, and tie with string around meat end.

5. Stand tied chops, bone side down, on a very lightly oiled baking sheet. Sprinkle lightly with pepper.

6. Bake in the middle of a preheated 425° oven for approximately 20 minutes for medium-rare chops.

7. Garnish with watercress bouquets and lemon wedges and serve accompanied by baked tomato halves Provençal.

PROGRAM DAY SIX

Tomatoes Provençal

42 Calories • 5.5 gr. Carbohydrates • 1.3 mg. Sodium

INGREDIENTS

2 firm, ripe tomatoes
2 Tb breadcrumbs made from gluten bread
1 tsp toasted wheat germ (optional)
1 to 1½ tsp minced garlic to taste
2 tsp minced fresh parsley
¼ tsp dried thyme or 1 tsp minced fresh thyme
1½ tsp pure olive or safflower oil
Freshly ground pepper to taste

METHOD

1. Preheat oven to 425°.

2. Cut tomatoes in half and carefully scoop out seeds with a small spoon. Place halves, cut side down, on a rack to drain.

3. Combine breadcrumbs, wheat germ, garlic, thyme, parsley, and oil in a small bowl.

4. Place tomato halves, cut side up, in a lightly oiled baking dish and top with breadcrumb mixture.

5. Bake in the middle of a preheated 425° oven for 15 minutes or until lightly browned.

6. Serve with lamb chops Provençal.

Cold Strawberry Yogurt Soufflé

115 Calories • 22.4 gr. Carbohydrates • 45.6 mg. Sodium

INGREDIENTS

1 3½-cup soufflé dish or 4 ½-cup individual soufflé dishes
Aluminum foil and string or tape to make collars

1 pt. fresh strawberries or 1¾ cups thawed whole frozen unsweetened strawberries
3 Tb fresh orange juice
1 envelope unflavored gelatin
3 Tb honey
½ tsp finely minced orange zest
¾ cup plain yogurt
2 large egg whites

Garnish:
1 small bunch fresh mint

METHOD

1. To make a collar for a soufflé dish, cut a piece of aluminum foil long enough to go around the dish, overlapping slightly, and wide enough when folded in half to stand 1 inch above the top of the dish. Tie or tape foil securely around the soufflé dish.

2. Reserving a few berries for garnish, purée the remainder.

3. Place orange juice and gelatin in a small bowl. When gelatin has absorbed the juice, place bowl in a pan of boiling water and stir mixture until it becomes liquid.

4. Stir honey and orange zest into gelatin mixture.

5. Thoroughly combine strawberry purée and honey gelatin mixture in a food processor or blender or by whipping with a metal whisk.

6. Add yogurt to mixture and gently combine. Be careful not to overblend, or yogurt will liquefy.

7. Beat egg whites until stiff but not dry.

8. Add ¼ of whites to strawberry mixture and fold in thoroughly with a plastic spatula.

9. Add remaining whites and fold in gently but thoroughly until no white streaks remain.

10. Pour or spoon mixture into soufflé dish or dishes and refrigerate for at least 2 hours or until mixture is firm.

11. Carefully remove collar and garnish soufflé with remaining strawberries and fresh mint.

I. LOTUS POSTURES
especially beneficial to thighs and hips

(a) HALF LOTUS POSTURE

Preparation for these postures is best achieved from the lying-down position. Bring left foot to the groin, trying to press left knee as close as possible to floor.

(b) Maintaining position a, lift arms overhead; hold 10 counts; reverse legs.

These half positions may eventually be executed from the sit-up position and will thereafter lead to the achievement of the Complete Lotus posture — both legs crossed.

(c) COMPLETE LOTUS POSTURE

the locked position of legs stimulates circulation in the abdominal region and encourages proper elimination

II. STANDING STRETCHES
help to stimulate cells and reduce fat around hips and thighs. They also strengthen legs.

(a) STANDING STRETCH

As you inhale, stretch one arm upward and the opposite arm downward; hold 4 counts, reverse.

PROGRAM DAY SIX

(b) STANDING BALANCE STRETCH

Stretch right arm overhead, take hold of left foot with left hand, bringing heel to buttock; hold 10 counts; reverse.

(c) STANDING TWIST STRETCH

With both hands stretched overhead, hands clasped, palms turned upward, slowly twist to the left, center, and right, holding 2 counts in each position; return to center and relax.

Program Day Seven

Menu

BREAKFAST

> ½ cup fresh strawberries or ¼ small cantaloupe with lime wedge
> Soft-boiled egg with asparagus tips
> 1 slice gluten toast (optional)
> ½ tsp whipped unsalted butter (optional)
> 1 tsp honey (optional)
> Herb tea

LUNCH

> Sole filets with julienne of vegetables baked in papillotes
> Ricotta pudding
> Herb tea

DINNER

> Gaspacho Andaluz (spicy cold tomato soup)
> Lemon chicken brochettes
> Brown rice pilaf
> Oranges with cinnamon
> Herb tea

Exercises

- STANDING CHEST EXPANSION
- TRIANGLE TWIST
- BACK BEND AND TWIST
- HIP BALANCE

PROGRAM DAY SEVEN

Soft-Boiled Egg with Asparagus Tips

SERVES 1

117 CALORIES • 7.3 GR. CARBOHYDRATES • 73 MG. SODIUM

INGREDIENTS

¼ lb. thin asparagus spears (stalks should be firm
and crisp with tightly closed tips)
1 egg
Pepper and paprika to taste

METHOD

1. Place one hand at each end of an asparagus stalk and bend the stalk until it breaks in two. It will break at the point where the tender tip ends and the tougher part of the stalk begins.

2. Scrape off the scales on the lower part of tips and wash tips in cold water to remove any sand. The butt ends may be reserved, then peeled and cooked for use in soups or purées.

3. Steam tips in a vegetable steamer for 1 minute or less. They should remain crisp. Plunge into cold water for a few seconds to preserve their fresh color.

4. To assure that the whites of soft-boiled eggs will remain tender, it is best to start the cooking in cold water. Place egg in a small saucepan with cold water to cover.

5. Heat the water gradually to a simmer but do not let it boil. When water begins to simmer, cover the pan and remove it from heat. Let the egg stand for 2 minutes in the covered pan for a 3½-minute egg or 3 minutes for a 4-minute egg.

6. Place egg in an egg cup and remove its top. Arrange the egg cup and asparagus tips decoratively on a plate. Sprinkle the egg with pepper and paprika and dip the asparagus tips into the egg as a pleasant alternative to toast.

PROGRAM DAY SEVEN

Sole Filets with Julienne of Vegetables Baked in Papillotes

210 CALORIES • 9.35 GR. CARBOHYDRATES • 208.0 MG. SODIUM

INGREDIENTS

4 6-oz. filets of sole, flounder, or other white fish

MARINADE:
4 Tb wine vinegar
4 tsp safflower oil
4 tsp minced shallot or green onion
4 tsp fresh minced mint or 1 tsp dried
4 tsp fresh minced parsley
2 tsp fresh minced tarragon or ½ tsp dried

2 medium-size carrots, peeled
1 medium-size white turnip, peeled
1 medium-size zucchini, well scrubbed
4 sheets of cooking parchment paper or aluminum foil, 10" x 12"
1 to 1½ tsp safflower oil
1 Tb Dijon mustard

GARNISH:
Small bouquets of watercress or parsley
4 lemon or lime wedges

METHOD

1. Rinse filets under cold running water and pat dry with paper towels.

2. Combine marinade ingredients. Place filets in a shallow glass or earthenware dish and pour marinade over them, making sure they are well coated.

3. Marinate for at least ½ hour at room temperature, or for 1 hour or more refrigerated, turning filets from time to time.

4. Cut vegetables into matchstick-like julienne strips, approximately 1½" in length. This is most easily accomplished by cutting carrots into 1½" lengths and cutting those sections into strips. Turnips and zucchini should first be sliced into rounds ⅛" thick, the rounds cut into julienne strips.

5. Fold sheets of parchment or foil in half across the width, making rectangles 5" by 12". Starting at the center crease, cut rectangle into a half-heart shape so that when opened, it resembles a heart.

6. Preheat oven to 375°.

7. If using cooking parchment, lightly oil both sides. If using foil, oil only the side that will touch fish.

8. Place a sole filet on an opened heart shape next to the center crease. Arrange one quarter of vegetable julienne decoratively on top of filet and top with one quarter of marinade.

9. Fold the other half of heart over fish and vegetables. Starting at the top of center crease, fold edges together in a pleating motion to enclose filet in a sealed packet or papillote. Repeat this operation with remaining filets.

10. Place papillotes on a baking sheet and bake in the mid-

dle of a preheated oven for 10 to 12 minutes. Papillotes should be puffed and brown, fish flaky, and vegetables crisp.

11. Traditionally, papillotes are served directly on garnished plates, but if using foil, you may wish to carefully remove fish before serving to avoid a "TV dinner" appearance.

Ricotta Pudding

128 Calories • 7.8 gr. Carbohydrates • 128.0 mg. Sodium

INGREDIENTS

1 cup whole-milk ricotta
1 Tb honey
¼ tsp vanilla
2 egg whites
Cinnamon to taste
¼ cup fresh strawberries or raspberries (optional garnish)

METHOD

1. Combine ricotta, honey, and vanilla in a small heavy-bottomed saucepan or double boiler.

2. Beat 2 egg whites until stiff but not dry.

3. Heat ricotta mixture, stirring with a wire whisk until hot but not yet simmering.

4. Pour ricotta mixture into a medium-size mixing bowl and carefully fold in the egg whites one half at a time, using a rubber or wooden spatula.

5. Spoon pudding into 4 wine glasses or pudding dishes. Sprinkle with cinnamon.

6. Chill pudding uncovered in refrigerator until ready to serve. Garnish with fresh berries.

NOTE: Pudding may begin to separate if made more than a few hours in advance.

PROGRAM DAY SEVEN

Gaspacho Andaluz
(Spicy Cold Tomato Soup)

142 CALORIES • 17.4 GR. CARBOHYDRATES • 6.8 MG. SODIUM

INGREDIENTS

4 Tb breadcrumbs made from gluten bread (no crust)
2½ tsp minced garlic
3 Tb wine vinegar
2 Tb good-quality olive oil or safflower oil (preferably cold pressed)
6 ripe tomatoes (approximately 2½ lb.), blanched, peeled, seeded, and chopped
1¼ cups chopped, peeled, and seeded cucumber
1¼ cups coarsely chopped, seeded, and ribbed green pepper
3 Tb coarsely chopped onion
Freshly ground pepper to taste
½ tsp tomato paste (optional)

METHOD

1. Combine breadcrumbs, garlic, vinegar, and oil in a 1-qt. mixing bowl.

2. Reserve ¼ cup chopped tomato, ¼ cup chopped cucumber, and ¼ cup chopped green pepper.

3. Purée remaining vegetables with breadcrumb-garlic mixture in a food processor, blender, or foodmill. The purée should be pink in color and have a smooth, creamy texture.

4. If texture is not smooth enough, press purée through a wire-mesh strainer, using the back of a wooden spoon. If tomatoes were pale, you may wish to add ½ tsp tomato paste to mixture for richer color.

5. Season soup with freshly ground black pepper.

6. Serve gaspacho well chilled, garnished with reserved chopped vegetables.

PROGRAM DAY SEVEN

Lemon Chicken Brochettes

257 CALORIES • 8.75 GR. CARBOHYDRATES • 74.6 MG. SODIUM

INGREDIENTS

2 large skinned and boned chicken breasts, approximately 2 lb. weighed unboned (reserve skin and bones for low-sodium chicken broth)

MARINADE:
4 Tb lemon juice
2 Tb safflower or olive oil
½ tsp ground ginger
½ tsp ground cumin
Black pepper and paprika to taste
4 small onions quartered

4 large metal skewers or wooden skewers soaked in water

GARNISH:
4 lemon wedges
2 tsp minced parsley

METHOD

1. Cut chicken breasts into bite-sized pieces.

2. Combine marinade ingredients in a shallow bowl. Add

chicken and onion quarters and marinate for 30 minutes to 1 hour at room temperature or for several hours refrigerated.

3. Thread pieces of chicken and onion on skewers. Sprinkle with paprika and pepper to taste.

4. Grill brochettes in a preheated broiler or on a charcoal grill for 3 to 4 minutes on each side.

5. Serve brochettes with brown rice pilaf. Garnish with lemon wedges and minced parsley.

PROGRAM DAY SEVEN

Brown Rice Pilaf

SERVES 6

175 CALORIES • 28.8 GR. CARBOHYDRATES • 4.45 MG. SODIUM

INGREDIENTS

1 cup homemade chicken broth (see recipe page 22) or low-sodium canned
1 cup unsweetened apple juice
2 tsp apple-cider vinegar
1 tsp safflower oil
4 Tb minced onion
1 cup brown rice
2 Tb raisins
¼ tsp allspice
¼ tsp cinnamon
¼ tsp turmeric
¼ tsp ground ginger
⅛ tsp black pepper or to taste
1 Tb minced parsley

METHOD

1. Combine chicken broth, apple juice, and vinegar in a small heavy saucepan with lid. Bring to a boil.

2. Heat oil in a non-stick skillet; add onion and rice and sauté briefly.

3. Add onion and rice to boiling broth mixture and stir in raisins and spices.

4. Cover and reduce heat. Cook over low heat for 45 minutes or until liquid is absorbed and rice is tender.

5. Fluff lightly with a fork to release steam and prevent overcooking of bottom.

6. Serve as an accompaniment to fish or fowl.

PROGRAM DAY SEVEN

Oranges with Cinnamon

78 CALORIES • 20.0 GR. CARBOHYDRATES • 2 MG. SODIUM

INGREDIENTS

3 large navel oranges
1 tsp lemon juice
Cinnamon to taste

METHOD

1. Peel oranges, removing all white inner skin.

2. Holding a peeled orange over a bowl to catch juice, cut

between section membranes with a small sharp knife and let sections fall into bowl.

3. Squeeze the remaining membranes to remove any juice and discard. Continue in the same manner with remaining oranges.

4. Add 1 tsp lemon juice to orange sections and sprinkle with cinnamon to taste.

5. Check for sweetness and add honey (61 calories, 16.5 gr. carbohydrates, 1 mg. sodium per Tb) to taste if necessary.

6. Serve orange sections in chilled stemmed wine glasses.

I. STANDING CHEST EXPANSION
for back, arms, abdomen, and legs

Start from a standing position, hands clasped behind at lower back (not shown).

(a) With feet close together, slowly bend forward, squeezing arms as high as possible behind your back and bringing head to your knees; hold 10 counts, return to standing position.

(b) STANDING CHEST EXPANSION VARIATION
From the open-leg position bend forward to right knee; center; left knee and center; hold each 4 counts.

II. TRIANGLE TWIST
for hips, thighs, and waist

From the open-leg position, arms outstretched, slowly twist to the left, bringing the right hand to the left foot; look up at the left hand; hold 6 counts, reverse.

III. BACK BEND AND TWIST
firms buttocks, thighs, and lower back muscles

(a) One knee up, toes on floor, opposite leg extended with knee on floor; inhale, stretching arms upward; exhale, arching back. Hold 6 counts.

(b) Arms overhead, fingertips touching, twist to right, look back over shoulder; hold 6 counts, reverse.

IV. HIP BALANCE
stretches hamstrings, neck, and shoulders; strengthens the abdominal muscles.

Starting from the lying-down position, arms at side, bring knees to chest, extending legs straight up; slowly lift head off floor and raise arms upward clasping your toes; hold 6 counts. Release by bending knees and return to starting position.

Index

RECIPES

BREAD
Gluten 24
Triticale 24

CUSTARD AND PUDDING
Louisiana Honey 64
Ricotta 145

EGGS
Baked
  Mexican-Style 98
  in Potatoes (Parmentière) 120
Omelets
  with Herbs 28
  Puffy Asparagus and Cheese 62
Soft-Boiled
  with Asparagus Tips 140

FISH
Filets in Papillotes 142
Salmon Steaks 34
Steaks Moroccan-Style 86

FRUIT
Apple
  Purée 106
Avocado
  Exotic Fruit Salad 100
Banana
  Energy Drink 60
  Sherbet 102
Melon Swan 124
Oranges with Cinnamon 152
Papaya
  Boats 88
  Exotic Fruit Salad 100
Peach
  Energy Drink 60
  Frozen Yogurt Melba 47
Pears
  Steamed 51
Pineapple
  Sherbet 72
Strawberry (ies)
  Ice 36
  in Orange Baskets 82
  Yogurt Soufflé 132

ICES AND SHERBET
Banana 102
Pineapple 72
Strawberry 36

LEGUMES
Chickpeas
  in Harira Soup 30
Lentils
  in Harira Soup 30
  with Rice (Kitcheri) 70

MEAT
Beef
  Kabobs 49
Chicken Breasts
  Brochettes 148
  with Sweet and Sour Sauce 106
  Tandoori 68

INDEX

Lamb
Chops Provençal 128
in Harira Soup 30

RICE
with Lentils (Kitcheri) 70
Pilaf 150

SALADS
Carrot and Orange Juice 29
Exotic Fruit 100
Fennel and Arugula 33
Green Bean and Onion 104
Mushrooms, Watercress, and
　　Endive 44
Orange and Onion 61

SAUCES
Carrot 49
Herbed Yogurt Hollandaise 120
Peanut 49
Sweet and Sour 106
Vinaigrette 100, 126
Yogurt Bernáise 34

SOUFFLÉS
Cheese and Herb 45
Cold Strawberry Yogurt 132
Lemon 109

SOUPS *(Cold)*
Cucumber-Yogurt 84
Curried Fruit 66
Spicy Tomato (Gaspacho
　　Andaluz) 146

SOUPS *(Hot)*
Chicken Broth 22
Chicken and Lime 80
Hearty Moroccan-Style (Harira)
　　30
Turkey Broth 22
Watercress 48

VEGETABLES
Asparagus
　　and Cheese Omelet 62
　　with Soft-Boiled Egg 140
　　with Vinaigrette Sauce 126
Cucumbers
　　Cold Yogurt Soup 84
Tomatoes
　　Provençal 130

YOGURT
with Cinnamon and Honey 32
Cold Strawberry Soufflé 132
Frozen Peach Melba 47

EXERCISES (1)

Back Bend and Twist 156
Back Leg Lifts 76, 92
Ball-up Position 41
Bow Posture 112
Camel Posture 116
Cat Hump Resistance Stretch
　　114
Cobra Position 111
Forward Stretch 94
Frog Balance 56
Hip Balance 157
Knee to Chest Position 38
Knee Press-downs 93
Leg Lift-ups 39

Lotus Postures 134
Off-balance Sitting Position 54
Pelvic Lift-up 40
Pelvic Lift-ups 75
Plow Posture 117
Shoulder Stand 74
Side Leg Lifts 90
Spine Rock 53
Standing Chest Expansion 154
Standing Stretches 136
Tent Position 57
Triangle Twist 155
V Posture 113

INDEX

EXERCISES (2)

ABDOMEN
and legs
  Knee to Chest Position 38
legs, and back
  Ball-up Position 41
  Forward Stretch 94
  Leg Lift-ups 39
legs, back, and arms
  Standing Chest Expansion 154
and thighs
  Camel Posture 116
thighs, and back
  Pelvic Lift-up 40
and spine
  Cat Hump Resistance Stretch 114

CIRCULATION
Complete Lotus Posture 134
Shoulder Stand 74
Spine Rock 53

FLEXIBILITY
hip joint
  V Posture 113
pelvis
  Knee Press-downs 93

REDUCING
abdomen, hips, and thighs
  Bow Posture 112
  Standing Stretches 136

SHOULDERS
arms, abdomen, legs, and back
  Standing Chest Expansion 154
neck, hamstrings, and abdomen
  Hip Balance 157
release of tension
  Bow Posture 112
  Standing Stretches 136

SPINE
and abdomen
  Cat Hump Resistance Stretch 114
and back
  Cobra Position 111
legs, and circulation
  Complete Lotus Posture 134
  Spine Rock 53

THIGHS
back of thighs
  Tent Position 57
inner thighs
  Knee Press-downs 93
outer thighs
  V Posture 113
and abdomen
  Camel Posture 116
and back
  Plow Posture 117
back, and abdomen
  Pelvic Lift-up 40
and buttocks
  Back Leg Lifts 76, 92
  Off-balance Sitting Position 54
  Pelvic Lift-ups 75
  Side Leg Lifts 90
buttocks, and lower back
  Back Bend and Twist 156
and hips
  Lotus Postures 134
hips, and waist
  Triangle Twist 155
legs, and feet
  Frog Balance 56
  Tent Position 57